Climate and Much W Dangers We Ignore

Howard Johnson

Climate Change Commentary from 2005 to the present
Are we actually heading into the next ice age?

The Cover

The photo on the cover shows the sun rising over ice-age Michigan. The ice is a mile deep and moving south toward Indiana. The weight of this ice has depressed the land, is grinding everything beneath it into boulders of many sizes, gravel, sand, and even down to particles of fine mud as it moves. It also scours any rock formation with which it comes in contact, often leaving long, parallel gouges. The front of the ice flow acts like a bulldozer, pushing up mounds of earth and rocks and gouging out holes that will become lakes and ponds when the ice melts. These mounds called glacial moraines are leveled, distorted and piled on top of each other as the leading edge of the ice advances and retreats repeatedly over decades and centuries.

Water flowing through and beneath the ice sculpts the land as well, eroding paths in the land which become rivers, streams, lakes, and ponds. This happens when the ice melts during the summer and as the planet warms. When the ice finally melted away completely, the land, freed of the weight of the millions of tons of ice, began a slow rise that continues to this day. Left behind near the high point of the moraine were huge deposits of sand, gravel, rocks, and boulders. Deposits of fine sand were left behind at the southern tip of Lake Michigan forming the landscape that is now Indiana Dunes State Park. Today sand and gravel are mined from deposits called gravel pits and used as building materiels, concrete in particular. These deposits are scattered throughout the areas affected by the ice sheet. Around many farms in the areas covered by the glacial ice, there are piles of glacial boulders removed from fields converted to farmland. Many of these boulders are from rock formations hundreds of miles to the north in Canada. Boulder removal goes on to this day in many farms.

The photo is quite obviously not of what is described, though it is an accurate representation of Michigan twelve thousand years ago. This photo was taken of a sunrise over the ice cap of Greenland where the ice actually is a mile deep and moving south. Maybe in a few millennia this will also be farmland if the planet warms up that much.

PREFACE

This is for those brave souls who read this book. This includes those from my friends and family on the far right who see me as a flaming liberal, and the similar group on the left who see me as a right-wing, fundamentalist. The reasonable and open-minded individuals who are not members of either extremes need no explanation as they will read it with a far more open and appreciative mind. Many of these reside in what Washington insiders, and the elite ruling class regardless of where they live, like to call flyover country.

Belief in *anthropogenic global warming*, or *climate change* as it is now called by the PC crowd, is far more a political movement than a scientifically proven reality. Any discussion or reporting on anything related to this movement must take into account the politics involved. For this reason, this book contains considerable examination of the political forces at work in what would otherwise be a straight forward scientific report. My apologies to those offended by this, but it is reality.

All of the contents, except for the quotes noted, are written as guided by my own opinions. I will therefore provide you with some declarations of where I get the basis for my opinions and my political reasoning. It explains how I see myself and how to understand my words. A personal statement:

I am a believer in myself and those individuals I trust.

I trust no politician, political operative, political activist, government official, celebrity, or media reporter or talking head I do not know personally, and very few of those I do. I trust no Muslim, ever. Their religious principles make this imperative.

The current actions of their extremists, the hatred and especially their murders of those who do not follow their religion, the silence and even support so called moderate Muslims continue to express in response to the atrocities and evil actions of their extremists are completely unacceptable in my eyes.

I trust and admire the rational opinions and logical judgements of the so-called **common** *people*. Their wisdom is far greater than that for which they are given credit. However, I do not trust their opinions or judgements when based on emotions, as they are far too easily swayed by the false, often persuasive words of those described in the previous paragraphs.

Far too often I hear people aping the words they have heard over the media, mostly the dictates of the New York Times. They have let the media do their thinking and defining for them. I expect that from liberals and all those on the left because that is just who they are, mental lemmings blindly following the crowd and rarely thinking for themselves. Unfortunately, I often hear the same type of things said by many of those who overtly oppose liberalism. They too have let the media mislead their thinking and define people and situations for them.

I do not expect anyone to understand any of that which they do not know. I do not even expect them to understand much of that which they do know.

I see politics, religion, pseudo science, culture and many other similar systems as powerful belief systems often used by unscrupulous individuals to control others for their own purpose.

I am not a follower of or beholden to any ?ism, ?ist, group belief system (religious, political, cultural, or other), political party, union, peer group, grant committee, dean or head of faculty, political or other boss, or corporate officer at any level. This is why I am free to express my own opinions without disrespect, concern for, or apologies to anyone or any group. I will change my own beliefs to fit new realities and knowledge when and if the new information or understanding requires it.

I consider myself a truly independent and quite liberal individual, not a progressive or regressive, but a realist who knows what it means to conserve, an equal opportunity supporter or detractor. Realism and being a realist are the only *?ism* or *?ist* I believe in and best describes my political philosophy. Because of this, I know my words and opinions may offend some individuals, but it is not my intent to do so. There are exceptions, of course.

I resent the current "Liberals" redefining the term to mean almost the opposite of its original meaning. The original "Liberals" were champions of small government, fiscal responsibility and individual rights. They have abandoned this philosophy and become part of the "Progressive" or "Communist" movement which espouses government control of virtually every facet of life for all individuals and seeks to destroy free enterprise in America and replace it with a "Nanny" state. They should more properly be called "Regressives."

I am not ever in any way controlled, intimidated or cowed by any kind of political correctness. I believe it to be a controlling creation of the many narcissist members of the entertainment world and in particular the TV news media. These self-serving hypocrites use **political correctness** to coerce people into speaking and thinking the way they determine. It is merely one more system that elitists use to try to control others, mostly the gullible, unthinking lemmings so many people, including Americans have become.

I will not accept as a fact, any words, concepts or ideas that do not meet the tests of logic, reason and/or hard science as I understand them. My opinions and beliefs are subject to change when and if new information makes a change necessary. I see the inflexible, closed mind - the mind of the fundamentalist of any flavor, religious, political (left, right, or in the middle), social, cultural, or other - as an evil curse on the individual whose mind is closed or inflexible for any reason

I no longer have patience for certain things, mostly personal things, not because I've become arrogant, but simply because I reached a point in my life where I do not want to waste time with what displeases me or hurts me. I lost the will to please those who do not like me, to love those who do not love me, to smile at those who do not want to smile at me, and be around those who cause me pain. I have a strong will to please those who like me, love those who love me, smile at all but those detractors mentioned earlier, and be around those who bring me

joy. I no longer spend a single minute on those who lie, fake affection, or want to manipulate. I decided not to coexist with pretense, hypocrisy, dishonesty, political correctness and cheap praise anymore. I have no patience for cynicism, excessive criticism and demands of any nature. However, I do solicit and encourage constructive criticism of my professional actions, especially my writing. I seek that which will improve and add value to my work

It is clear to me that thousands of free and independent individuals and groups working in a favorable competitive environment, under a capitalist system with limited government in a democratic republic, are infinitely superior in virtually every way to a central decision making collectivist body or government of any kind. The bigger and more powerful the government, the less freedom individuals have to grow and improve their life and the lower will be the standard of living under that government. Examples of this reality abound now and throughout history for at least the last 3,000 years. Freedom works. Collectivism works only for the few who are running the system and eventually will lead to impoverishment, dependency, and ultimately dictatorship and some form of slavery for the masses.

I believe in treating every individual with respect and honesty. These both deserve respect and honesty in return. However, I see no reason to be bound to do the same when faced with disrespect or dishonesty. I will always expect respect and honesty first.

I try to deal with every person with consideration in all of these things.

—Howard Johnson - 2012

"There are many who find a good alibi far more attractive than an achievement. For an achievement does not settle anything permanently. We still have to prove our worth anew each day: we have to prove that we are as good today as we were yesterday. But when we have a valid alibi for not achieving anything we are fixed, so to speak, for life. Moreover, when we have an alibi for not writing a book, painting a picture and so on, we have an alibi for not writing the greatest book and not painting the greatest picture. Small wonder that the effort expended and the punishment endured in obtaining a good alibi often exceed the effort and grief requisite for the attainment of a most marked achievement."

—*Eric Hoffer*

The emotional ranting of all politicians I consider as nonsense no matter how close to my own beliefs their words are sheeted. I have found that the real truth almost always lies in the opposite direction from where they are pointing. This begs the question, "How do you know when a politician is lying?" The answer, given many times by many people is, "When his or her lips are moving."

The highly biased media are a force to be reckoned with that is slightly above politicians in their dishonesty. They are far more insidious because they claim, and have cleverly convinced many to believe, that they are simply and objectively reporting the truth. Since most Americans still get the preponderance of their news from the so called *main stream* media, the leftist talking heads that populate the news reporting, have been a major factor in the promotion of the far left and the degradation of capitalism, freedom, independence, and conservative ideas and people.

Bernie Goldberg in his best-selling book, *Bias*, and John Stossel in several of his best sellers along with others wrote extensively about this devious, fraudulent practice in many popular books. I'm quite certain no self respecting liberal Democrat would read any of these books. Each author was a previous member of the far left who learned and reported these goings on in the media and were excoriated by their colleagues for revealing these truths.

Most members of the media have succumbed to the enticements of the entertainment world and have joined other entertainment celebrities in promoting the political far left. The Obamas are their present royal couple about whom no ill is ever spoken. The Obamas have become the American equivalent of Ferdinand and Imelda Marcos of the 1970s Philippines as far as media reporting is concerned. (I wonder how many pairs of shoes Michelle has?) It is hard to detect media bias as most of their effort is in **not** reporting negative

actions of Democrats and their far left members and supporters, while front paging any similar actions of Republicans and conservatives. Examples are legion including literally thousands of times when the New York Times headlines a questionable story regarding the right, then prints a correction or revision in small print hidden far back in the bowels of the paper. Their rule book seems to tell their reporters, "Write no ill about liberal Democrats or their actions, use only positive slants, but amplify any possible ill that can be found or implied about conservatives or Republicans and their actions."

One example of media silence about a very serious change in Social Security involves the changes made by Bill Clinton and his first Democrat controlled Congress. Democrats lead by Clinton removed food and several other important items with volatile pricing from the calculation of the COLA. By doing so, Democrats decimated many of the elderly when these changes they made reduced current SS payments to about half of what they would have been if these items were included in the COLA. This factor is used to adjust SS payments, for changes in the cost of living. The media has completely and deliberately ignored this assault on seniors by Democrats. Meanwhile, they keep saying "Republicans want to reduce your SS payments." The truth is Democrats are the only ones who have ever cut SS payments and they cut them drastically. Apparently Democrats believe seniors don't need food or the other items. There should be media outrage about this huge assault on seniors, but not a peep has ever been heard from the *compassionate* media. I wonder why the public doesn't know about this?

When Sarah Palin became the Republican vice-presidential candidate, the media dispatched hundreds of investigators to Alaska to dig up or fabricate as much dirt as they could imagine to use in their *news* broadcasts. They and their Hollywood friends mounted cruel personal attacks, ridiculing this lady for everything, mostly invented by their minions. No mention was ever made of her substantial accomplishments as governor in turning around the financial condition of the state even against the objections of numbers of Republican politicians there.

The media and anthropogenic global warming, AGW: The main stream media—referred to variously as the leftist media, the federal government's propaganda machine, the socialist press, or any number of quite accurate descriptive phrases—accepts and promotes AGW as a proven scientific fact. It is not. There are at least 17 other natural phenomena, known to affect weather and climate far more than the minuscule effect of changes in the percentage of CO_2 in the atmosphere. A number of these including a newly discovered one are listed in this book. See page 51. The AGW furor is much more a political movement, almost a religion, than a scientific theory. It is long on emotional appeals and scare tactics and very short on hard science.

One of the most comprehensive articles refuting the claims of Michael Mann and his followers with sound scientific questions and skepticism was published on the Internet recently. Here's where this science-based, in-depth article can be found:

http://www.john-daly.com/hockey/hockey.htm. The article is titled. *The `Hockey Stick': - A New Low in Climate Science* - by John L. Daly.

I recommend everyone who has any question at all about the efficacy of climate science read this article. It certainly reasserts the scientific skepticism that prompted this little book. It also reaffirms the author's concern, doubts, and comments expressed in many different sections. The fact is that one of the most powerful principles of good science is healthy and in depth skepticism and the questions prompted by that basic tenet of science.

A NOTE from Howard Johnson: Most members of the church of Anthropogenic Global Warming (AGW) treat any and all persons who do not completely accept their definition of the causes of global warming, (human activities, and in particular,

the use of fossil fuels) as **deniers** of global warming. They lump all who don't agree with them into the same indiscriminate category of **deniers** of AGW. This is done regardless of how accurate, intelligent, or scientifically based are their thoughts, ideas, or comments. The minds of the fundamentalist members of the church of AGW are quite closed. You cannot have an intelligent conversation or exchange of ideas with a person whose mind is closed and the AGW crowd certainly have closed minds.

One of the basic tenets of good science is an open and inquisitive mind. Need I say more? Climate is always changing for an infinite number of reasons. I most certainly do not deny the possibility that the earth may be warming. However, the prevailing long range predictions have been and remain for a cooling trend. While it is far too short to be anything but an anecdotal indicator, the last few years since 1996 *(actually as far back as 1961)* have seen a lowering trend in average temperatures on the planet. I also do not deny the possibility that it may actually be cooling long range. I might even agree that man's effects on the environment could be a contributing factor. What I do not agree with is that the increase in atmospheric CO_2 from man's use of fossil fuels is the main cause of warming. There are too many other factors proven to have a much larger effect on global temperatures than CO_2 from fossil fuels. Also, there are too many systems adding CO_2 to the atmosphere besides use of fossil fuels. A number of these are described in this book. Personally, I believe the emotional adherents to the idea of AGW are driven by the vast sums of money that is involved. Political motives are powerful forces, particularly where money is concerned and AGW provides a vehicle for tons of money for politicians and academics who join and drive the church of AGW. In my opinion, that is the belief system and driving force behind AGW, not rational scientific ideas and conclusions or accurate data.

Some Conventions Used in Most All of My Writing

An explanation of the term FEUDALS: All intellectual/political groups on the left promote collectivist, totalitarian government in the long run. This government will control and manage the masses with growing power enforced with handcuffs, guns, and prison. These people and organizations go by the following names: liberal, progressive, socialist, communist, fascist, and many more that are less popular. Although I sometimes use one or more of these titles interchangeably, I have coined my own term, **FEUDALS** for any and all of them. **FEUDALS,** refers to one of the earlier manifestations of totalitarianism that first appeared in China more than 3,000 years ago. The system failed, particularly for the masses of common folk who suffered greatly both economically and physically. It was far better known and understood by Americans as the political system of the medieval European tribes, city states and nations.

Remember Robinhood, the champion of the people, and his nemesis, the Sheriff of Nottingham, the local despotic member of the elite, ruling class. The royals, the aristocracy, the lords and ladies of those times are fully repeated and reflected by the elitist members of our current government. Examples of the gross extravagance of the new royalty are easy to find and document. They include the Obamas, Nancy Pelosi, Barney Frank, Harry Reid, the Clintons, etc. Their gross extravagances include lavish trips with family and friends, all paid for by your federal tax dollars. How about the President's many golf games and the entourage that he takes with him?

Under the burden of the current ruling class in Washington, America has experienced by far the largest transfer of wealth from the middle classes to the elite, ruling class, the super wealthy, in our history. Whether by actual dollars or by percentage, it was the largest transfer of real wealth from the 99% to the 1% ever to have occurred in America. In spite of the major economic downturn, the afore mentioned individuals and their like continue to squander taxpayer money on lavish living. The media completely ignores this virtually criminal behavior—in liberals. Hmmmm?

Of course, the Republican house took away princess Nancy's private jet, but that's about all that has changed. Members of the elite ruling class in Congress continue to give themselves cushy retirement plans, expensive healthcare plans, and their travel and other perks, all paid for with taxpayer dollars extracted from the private sector in almost invisible taxes. Members of the ruling class, and even the workers in the bloated ranks of government bureaucracies, have not suffered any loss of income or been denied any of their lavish perks in the recession. At the same time, the working people, the middle class, most of the private sector individuals, have lost much of their wealth, their jobs, their homes, and their businesses. How long will the American people put up with a government that is squandering their future, as well as that of their children and grandchildren. These extravagances at the expense of the masses are in addition to the lavish, under the table perks provided by lobbyists and others seeking favors While Republicans participate in this looting of our treasury and benefit from the largess of lobbyists, they are rank amateurs compared with Democrats backed and supported by the media and millions of government bureaucrats.

Did you ever wonder how those in Washington, mostly Democrats, can arrive on the scene with such modest means and soon become billionaires, or at least, multi millionaires? Senator Harry Reid is a classic example. Or how about those super wealthy political families and dynasties. Did you ever wonder where the many billions in tax payer dollars that were spent on that huge Boston boondoggle, the *Big Dig*, ended up? What part of that money found its way into the coffers of the Kennedy family, conveniently and secretly moved there by contractors and unions involved in the project?

Did you ever wonder about WV Senator Robert Carlyle Byrd, an active member of the Ku Klux Klan, who held the titles Kleagle (recruiter) and Exalted Cyclops? When it came time to elect the *Exalted Cyclops*, the top officer in the local Klan unit, Byrd won unanimously. He was an active segregationist and, gave many anti black speeches over many years. In numerous speeches he described Dr. Martin Luther King, Jr. as "one of the most dangerous men in America" even after King was murdered. A Senator from 1959 to 2010, Byrd earned the title, "champion of pork barrel spending." Everywhere one turns in West Virginia are signs saying, "The Robert Byrd - - - -." These monuments to his use of taxpayers dollars to feather his political nest and satisfy his monster ego are all over the state. A political opportunist, he frequently changed his positions, most notably by renouncing his long time segregationist position late in his career to fit the times (and get elected). His ideals and principles were firmly planted wherever the political winds blew them. Such are those held in high esteem by liberal Democrats.

The recent home mortgage debacle was engineered by unelected liberal members of the monstrous Democrat bureaucracy who pocketed hundreds of millions in the process. Noting the obvious agenda of the Feudals, I find it hard to believe these actions were not deliberately aimed at destroying the wealth of middle class Americans to make them dependent on government. In the process, Jim Johnson and Franklin Raines, along with many others, were paid hundreds of millions in bonuses while they dismantled the controls on Wall Street banks, controls designed to prevent such a debacle. They were aided and abetted by people like those who founded Countrywide, almost all Democrat supporters. They all made hundreds of millions with the support of Democrats like Barney Frank, Chris Dodd and then Senator Obama. Taxpayers and people with mortgages paid dearly when the house of cards folded in 2007 and 2008. In the process, they destroyed most of the wealth of the American middle class.

When a few Republicans including President Bush and Vice President Cheney tried to stop this madness and reinstate abandoned mortgage and banking rules, the main stream media joined Democrats and branded their efforts immoral, racist, anti poor and every other vituperative label they could come up with. This all reminded me of what Karl Marx (remember him) said was the secret to success for communism. Karl Marx wrote,

"A prosperous and well-informed middle class is the biggest deterrent to communism. Reduction of the middle class from independence to near poverty is a necessary part of any communist revolution."

He also wrote, "Without violence nothing is ever accomplished in history." Marx would have fit right in with the Obama administration, the ridiculous federal bureaucracy, and the new liberal Democrat party.

Actually, the battle between collectivism and individualism now raging in our country, has been going on for as long as men have gathered in groups. From families to tribes to city states to nations, collectivist cultures have been the rule, always to the advantage and wealth of a small group, the ruling classes. Eventually, each of these ruling classes took advantage of the masses, the so-called common man, leading to the control, the impoverishment and often the enslavement of these people. Oligarchies of one type or another have been the most prevalent form of government down through the ages. Kings, emperors, sultans, pharaohs, chieftains, lords, commissars, priests, presidents, members of legislatures—the ruling elite in every group, have always used their minions to control commoners, serfs, subjects, and all lesser masses and individuals. In the most benevolent of these, loyalty to the ruler was maintained by peer pressure. This type of rule almost always gave way to tyrannical rule by a single ruler or group of rulers, who held privileges and control over lesser individuals, the common man, by coercion, fear and threat of imprisonment or death.

Eric Hoffer describes this type of rule as practiced by the communist government of the Soviet Union in his book, The Ordeal of Change, published in 1963. This is what he wrote:

"The Marxist movement has made giant strides during the past forty years. It has created powerful political parties in many countries, and it is in absolute power in the vast stretch of land between the Elbe and the China Sea. In Russia, China and adjacent smaller countries, the revolution envisioned by Marxism has been consummated. What, then, is the condition of their masses and the intellectuals in these countries?

"In no other social order, past or present, has the intellectual come so much into his own as in the Communist regimes. Never before has his superior status been so self-evident and his social usefulness been so unquestioned. The bureaucracy which manages and controls every field of activity is staffed by people who consider themselves intellectuals. Writers, poets, artists, scientists, professors, journalists, and others engaged in intellectual pursuits are accorded the high social status of superior civil servants. They are the aristocrats, the rich, the prominent, the indispensable, the pampered and petted. It is the wildest dream of the man of words come true.

"And what of the masses in this intellectual's paradise? They have found the intellectual the most formidable taskmaster in history. No other regime has treated the masses so callously as raw material to be experimented on and manipulated at will. Never before have so many lives been wasted so recklessly in war and in peace. On top of all this, the Communist intelligentsia has been using force in a wholly novel manner. The traditional master uses force to exact obedience and lets it go at that. Not so the intellectual. Because of his professed faith in the power of words and the irresistability of the truths which supposedly shape his course, he cannot be satisfied by mere obedience. He tries to obtain by force a response that is usually obtained by the most perfect persuasion, and he uses terror as a fearful instrument to extract faith and fervor from crushed souls."

—Eric Hoffer, The ordeal of Change, 1963

Thus I will apply the term **FEUDAL** to the elitist, intellectual despots that ignorant and subject people allow to rule over and control their lives. These cruel individuals and groups have been called by many names over the millennia: from scribes and pharisees, to kings and queens, to lords and ladies, to Presidents, Senators, Representatives and their spouses. I'm sure you see the similarities.

Of course there are exceptions. These are usually small countries with a common and homogenous language, culture, and racial makeup. Most are sustained financially by a substantial and valuable natural resource. The Scandinavian countries and especially Norway are the most prevalent examples.

REMARKS and NOTES: At various locations I have inserted my own comments among the writings of others. These are usually preceded by the words, REMARKS or NOTES, and are in *italics* or even ***bold italics*** to make it known they are my comments.

For example: I have expanded on a remark made by Thomas Sowell, to wit: *"The liberal Democrat party in the US has a record of hatred, failure, corruption, deception, and greed so blatant that only an intellectual could ignore or evade it."*

—*Howard Johnson, 2009*

Important note about the IPCC:

Most of the articles quoted in this book, indeed, all of those based on the IPCC (Intergovernmental Panel on Climate Change) information or by those who support the IPCC theory of anthropogenic global warming, are in lockstep. They make the **assumption** that increases in the amount of CO_2 in the atmosphere will have a substantial effect on increasing the air temperature, the retained solar heat, of the planet, and treat that as a proven and unassailable fact—a given by many people. In factual contrast, graphs of global temperatures plotted with the amount of atmospheric CO_2 over the last 415,000 years show clearly those changes in CO_2 **follow** changes in temperature. This means that changes in CO_2 are the **result** of temperature changes, not the **cause**. See the Vostok graph and explanation on page 15. **This fact alone indicates that AGW as promoted by the IPCC, is probably in error, that there are other factors not being considered in the computer models, factors like the effects of cosmic ray variations on cloud cover and the related reflection of heat from the sun.** *(see the section on Svensmark's theory of the effects of nearby super novas starting on page 16.)*

It's a sad state of affairs when a political movement takes over a theory difficult to prove or disprove and uses it to gain money and power. *Anthropogenic Global Warming*, has morphed into *Climate Change* for obvious reasons. It has become an emotionally charged movement where reason and the scientific method has been all but discarded. In this publication the author seeks to put a more realistic, rational face on the many theories involving Climate Change. Some far more dangerous menaces are also discussed.

Let's get some definitions straight. The terms *global warming* and *climate change* are all inclusive terms that have come to include all of the many various physical effects that contribute to changes in the climate of the planet. *Anthropogenic global warming, AGW,* is a very specific term that includes only those effects caused by human activities. These causes have come to include almost exclusively the increase in atmospheric carbon dioxide caused by human use of fossil fuels. Actually, the human destruction of forests may be a larger contributor of carbon dioxide to the atmosphere than the use of fossil fuels.

There are numerous effects that are demonstrably larger contributors to both the warming and the cooling of the planet than even doubling the amount of carbon dioxide in the atmosphere. Ten of the most well documented effects are listed on page 51 of this book.

Global warming may be a reality, but so is global cooling. Thousands of years of cyclical ice ages in the past provide evidence that ice ages will most likely come again. These cycles repeated many times long before man came on the scene. Extreme and dangerous anthropogenic global warming caused by man's emissions of CO_2 is not a rational reality. It is more likely a political fiction driven by greed and the lust for power. That fact certainly does not in any way lessen the major environmental problems, the havoc, explosive growth of the human population continues to wreak on the planet. There are several of these problems described in these pages.

Climate change is a reality we simply must deal with. We may even have the power, develop the technology, to modify climate. However, any change could have unforseen consequences. For example: one proposal to cut down on the sunlight and so to cool the earth involved the placement of billions of tiny reflective particles between the sun and the earth. The very real danger that such an action could tip the long range heat balance and plunge the earth into a major ice age. Such proposals should be given a great deal of study before even considering taking action. One tiny mistake and the results could be catastrophic.

One of the many articles I read stated that cloud cover would act as a blanket, retaining solar heat and warming the planet. In fact, the exact opposite is true. It is a well-proven and accepted fact that cloud tops reflect the sun's radiation and that increased cloud cover would lower the amount of radiation energy retained as heat in the atmosphere. This is one of the countless factors affecting the retention of heat by the earth. While our understanding of the effects of these complex, interacting systems continues to grow, our computer models are still poor predictors of hurricane seasons, one tiny part of the overall climate systems. Once we get models that accurately predict hurricane systems and their movements, then and only then will our models of worldwide climate even begin to be reliable.

CONTENTS

Graphics and Plots

INTRODUCTION

"Increasing numbers of scientists are joining me in taking the position that man's addition of carbon dioxide to the atmosphere has at most an infinitesimal negative effect on global temperatures, and that it does have a very positive effect of increased food production for a human world that desperately needs more food."

—Howard Johnson - April 2014

Here are a few background factors that determined my present position on the effects of increased atmospheric CO_2 on climate and in particular, global temperature variation, and on the human food supply. Why increasing numbers of scientists are joining me, and why I published this book.

In 1912 the meteorologist Alfred Wegener amply described what he called continental drift, expanded in his 1915 book The Origin of Continents and Oceans. It was a revolutionary concept and not well received by the scientific community. Then in the fifties and sixties and shredding the textbooks, Tuzo Wilson, Dan McKenzie and Jason Morgan merrily explained earthquakes, volcanoes, mountain-building, and even the varying depth of the ocean, simply by the drift of fragments of the lithosphere in various directions around the globe. The revolutionary theory of plate tectonics finally came on the scene. Starting when I was in college in 1945, I was fascinated and read everything I could find about this new theory. Then ensued a victory of the pioneers of plate tectonics in their battle against the most eminent geophysicists of the day. It took years of often contentious battles before the theory was given credibility by the scientific community. I remember at the time wondering why those who argued against it could not see its worth.

A mark of a good hypothesis is that it looks better and better as time passes. With the triumph of plate tectonics, diehard opponents were left red faced and blustering. In 1960 you'd not get a job in an American geology department if you believed in continental drift, but by 1970 you'd not get the job if you didn't. That's what a paradigm shift means. That leads me to another intriguing theory recently proposed (1996) that is currently being ignored by most and attacked by a few.

Plate tectonics was never much of a political issue, except in the Communist bloc. There, the immobility of continents was doctrinally imposed by the Soviet Academy of Sciences. An analogous diehard doctrine in climate physics went global two decades ago, when the Intergovernmental Panel on Climate Change was conceived to insist that natural causes of climate change are minor compared with human impacts. It is quite obvious that the growing "global warming" movement is not based on good science. Good science does not ignore data that disproves a hypothesis and look only at that which does. Good science does not ignore data points on graphs used to promote a concept just because those points do not agree with a particular agenda. Good science does not refuse to look at or consider basic math, chemistry, and physics which do not bear out the hypothesis. Good science does consider all the data, positive, negative, and anywhere in between. Good science does not place the opinions or theories of any person or group above the provable facts no matter what the reputation of the person or group. Good science does not completely ignore the fact that earth has now been in a cooling trend for the last sixteen years. Good scientists are very skeptical, especially so where politics or politicians are involved. It has been said and wisely so that one should not ask a scientist to disprove a paranormal claim, ask instead a magician.

I must say, I am not an expert on climate, but after at first jumping on the AGW bandwagon, I took a closer look at and studied data, in particular, the physics and thermodynamics of heat absorption and radiation by atmospheric gasses including carbon dioxide and water. I also referred to graphs and charts about climate, some going back more than four hundred thousand years. This information is from many sources and has the benefit of the test of years. To this discerning eye, the knowledge gained pointed to a major cooling event, possibly even the beginnings of another "Ice Age" in the very near future. These type events take place over millennia, even centuries, and can be the cause of huge changes in the ecology of much of the earth. The last ice age ended around ten thousand years ago after nearly sixty thousand years of holding most of the current temperate zones in an icy, arctic grip. North America was buried with up to a mile of ice as far south as the Ohio river and the Rockies were buried all the way to Mexico. Life in North America was tough for mammals and impossible for all cold blooded creatures. Trees and flowering plants were non existent anywhere north of the snow line. Europe and Asia were similarly affected with all mountains buried as well. Glaciers even flowed into the Mediterranean at times. Sea levels were much lower with land bridges connecting many places now under water. It was a very different world.

Many dramatic climate changes have occurred over very short time periods, decades or even less. In fact many scientists now believe sudden changes of a few decades or even a few years are the rule rather than the exception.

The Royal Astronomical Society in London published (online) Henrik Svensmark's paper entitled **"Evidence of nearby supernovae affecting life on Earth."** After years of effort, Svensmark shows how the variable frequency of stellar explosions not far from our planet has ruled over the changing fortunes of living things throughout the past half billion years. Appearing in Monthly Notices of the Royal Astronomical Society, It's a giant of a paper, with 22 figures, 30 equations and about 15,000 words. I was fascinated by the Svensmark hypothesis and the reporting on it by Nigel Calder. Nigel Calder, a respected British science writer, is the author of a marvelous science book, *The Magic Universe.* Calder reports on theories, observations and experiments, often long before publication. He writes, "Since 1996, Svensmark's hypothesis has brought new successes year by year and has resisted umpteen attempts to falsify or discredit it." Without going into details, it turns the tables on the IPCC claims of man's addition of CO_2 to the atmosphere as a catastrophic happening. This is covered in detail in this book, *Climate & Much Worse Dangers We Ignore.*

The universal counter argument I have heard literally hundreds of times to my opinions about AGW includes, "The Union of Concerned Scientists, or any other of dozens of similar names, warned us of the catastrophic effects of our adding CO_2 to the atmosphere by use of fossil fuels." Add to that the political forces at work and the financial bonanza they see from those carbon taxes, and a clear picture emerges. Then I ask the questions, "Why are there no arguments using the known physics, chemistry, and thermodynamics of atmospheric gasses? Where are the opposing theories about the effects of changes in atmospheric CO_2? Where are the studies of the beneficial results of increased atmospheric CO_2? Why is it there is virtually no reporting of any differing views?" Basically, the entire argument for AGW is, "We all say it is so and that makes it so."

Ask Dr. Judith Curry of Georgia Tech. As one of the eminent authorities on hurricanes and polar ice, Dr. Curry has been ridiculed by colleagues for not rejecting out of hand all comments of those they call "deniers" and for asking many of the same questions I ask. Remember, questioning is one of the most important tenets of the scientific method. I am reminded of one of my favorite quotes from Peter Abelard, "By doubting we are led to inquire. By inquiring we learn the truth." Incidentally, Abelard's words are a quite good description of the *scientific method.* Apparently the global warming crowd will tolerate no doubting or inquiring from anyone, even highly qualified scientists. I do not believe that is even a remotely scientific attitude. Good science welcomes doubters and questioners. As a matter of fact, the entire basis of the *scientific method*, is centered on repeated and thorough questioning. The idea that *consensus science* (a group opinion) is superior to *hard science* (math, physics, chemistry) is ludicrous. It goes against the grain of all true scientific reasoning, and in truth, **the opposite is always true. Hard science always trumps** *consensus science.*

A consensus means that everyone agrees to say collectively what no one believes individually.

—*Abba Eban (1915-2002)*

All truth passes through three stages. First it is ridiculed. Second, it is violently opposed. Third, it is accepted as being self-evident.

—*Arthur Schopenhauer*

So I say to all of you AGW catastrophist believers, "Show me the simple, basic facts, the physics, the chemistry, the thermodynamics, and the math that confirms your theory and I will gladly and willingly join your bandwagon. Until then, don't expect me to go along with the moving rush of lemmings because the PC crowd says I should."

The *Experts* Who Cry *Wolf* and are almost always *wrong*

The constant warnings of persistent attention-seeking intellectual doomsayers among, college professors, environmental extremists, PBS, the New York Times, Washington Post and other Newspapers, are becoming like the story of the boy who cried wolf. In their efforts to bring attention to themselves, gain celebrity, and sell themselves and their products to the public, these so-called pundits have exaggerated possible dangers out of all proportion to reality. These scare tactics sway a lot of people who are ignorant of the facts. They are so seldom correct in either action or scale that most reasonable people pay them scant attention anymore. The worrisome thing about this is that should a true dangerous menace come along (a real wolf) the predictions will be viewed as another exaggeration and ignored by the public.

EXAMPLE: When the very first primitive steam powered locomotives appeared at least one newspaper predicted people would die riding in these conveyances that could go as fast as forty miles per hour because they would be unable to breathe moving that fast. You may laugh, but how many erroneous and equally dire predictions have been foisted off on a gullible public by the media in the same way since then? There are several recent examples of this kind of activity recounted in this book. The media are quite obviously more interested in gaining attention for themselves than in telling the truth.

NOTE: Though I consider myself an environmentalist, I am not a blind believer in many of the extreme and ill-advised efforts of some so called environmental groups to stop all drilling for oil or gas, or mining for other essential materials. Time and time again, it has been proven that most if not all serious environmental damage can be prevented with proper, well-designed and engineered systems and enforcement of well-conceived laws to prevent such damage. Several examples of how numerous real and serious danger to the environment were countered and prevented by scientists and engineers are noted in this book. These effective, rational actions are almost never addressed or even mentioned by environmentalists who would shut down any and all projects with even minuscule risk of damage to the environment. They would not even let scientists and engineers apply solutions to the problems..

As PBS releases a special about the oil delivery system, the Business & Media Institute goes back in time to recall many environmentalists' pipe dream, stopping it.

* By R. Warren Anderson

* Wednesday, April 19, 2006 10:00 A.M. EDT

Alaska Pipeline Doom Sayings Revisited

After the discovery of oil in Prudhoe Bay, Alaska, it didn't take long for environmentalists to cry gloom and doom and for the media to hype those claims. From caribou dying to earthquakes to all hell breaking loose, there was no shortage of catastrophic predictions though the Alaska pipeline now boasts great success roughly 30 years later.

Construction on the pipeline began in 1975, and oil first moved through it on June 20, 1977. Former Secretary of the Interior Gale Norton summed up its success in 2003 that today the pipeline produces 17 percent of our domestic petroleum. It has pumped nearly 14 billion barrels of oil and $400 billion into our economy. We need the pipeline now even more now than when it was built.

In time for the PBS special, *The Alaska Pipeline*, set to air April 24 on PBS, the Business & Media Institute compared predictions from the pipeline's inception to the realities of the past three decades.

Read the following comparisons, the dire predictions and what actually happened, some reality checks.

Propaganda, Not Policy: Approval of the pipeline was not based on facts but on oil industry propaganda, according to some of the Department of Interiors top ecologists, reported The Washington Post on Feb. 11, 1971. On Nov. 14, 1973, the New York Times ran an editorial that began, "passage of the Alaska pipeline bill is the triumph of scare propaganda and economic pressure over reasoned public policy." These papers adhere to the leftist party line no matter how far from reality and truth it is. Shades of the old USSR.

Reality: Despite those claims, the pipeline has had tremendous policy implications. It created tens of thousands of jobs, from the construction of the pipeline in Alaska to the manufacturing of the pipe in Pennsylvania, to the building of the tankers to transport the oil to Louisiana.

And as gas prices head into another summer driving season, the pipeline's affect on the oil market bears mentioning. Alaska produces about 800,000 barrels a day or about 1 percent of the world market of 73.5 million barrels a day, said Peter Van Doren of the Cato Institute.

A loss of that production would increase prices by at least 10 to 16 percent. In the 1980s, when production was 1.8 million barrels a day and the world market was smaller (54 mbd), the loss of Alaskan oil would have increased world oil prices by 30 to 50 percent.

Bye-bye Caribou: Many people suddenly developed a passionate concern for the mating habits of Alaskan Caribou and campaigned noisily against intrusion of Arctic pipelines into this essential activity, reported The Christian Science Monitor on Oct. 10, 1972. The New York Times on Oct. 14, 1973, said the question is whether the caribou will go the way of the buffalo.

Reality: Thirty years later we can see the effects of the pipeline on the caribou. Walter Hickel, a former U.S. Secretary of the Interior and governor of Alaska, said that the caribou herd has not only survived, but flourished. In 1977, as the Prudhoe region started delivering oil to America's southern 48 states, the Central Arctic caribou herd numbered 6,000; it has since grown to 27,128. Alaska's Department of Fish and Game Web site reports that in general, caribou have not been adversely affected by human activities in Alaska. Pipelines and other manmade objects have been built to accommodate caribou movements, and the animals have adapted to people, machines, and the pipeline.

INSERTED NOTE: There have been many reports of caribou huddling near and even against the pipeline to stay warm in extremely cold weather and storms. It is quite possible that this could explain at least part of the herd's growth. So much for the doomsayer's predictions.

Earthquake Risk: Larry Moss of the Sierra Club stated in the Los Angeles Times on June 14, 1973, that the oil industry has continued, single-mindedly, its attempt to turn a sow's ear into a silk purse. Support for this claim was that the pipeline had basic design flaws which cannot really be overcome by engineering ingenuity. This was supposedly because the pipe would cross one of the most active earthquake zones in the world, would scar and despoil vast tracts of magnificent, undisturbed country and would threaten extensive oil spills in the numerous rivers which the pipeline would cross.

A report from top ecologists at the Department of the Interior claimed that dangers of severance in earthquake prone areas were inadequately dealt with, read The Washington Post on Feb. 11, 1971. The Alaskan area involved is renowned for its extreme seismic activity, the Post reiterated on May 7, 1972. In the 70 years before 1972, 23 major earthquakes had clobbered the terrain where the Alaskan pipeline would be built, any one of which could have caused a catastrophic break in the pipe, the Post article continued.

Reality: The time passed since the construction of the pipeline allows for testing of this claim. On November 3, 2002, a 7.9-magnitude earthquake struck Alaska. It was the worst earthquake recorded on Alaska's Denali fault, and considered a once-in-600-years event. The New York Times on Nov. 5, 2002, called it one of the largest earthquakes in American history, which, had it struck a major city, would have destroyed hundreds of buildings and killed many people. Tremors caused movements around Yellowstone National Park and even rocked boats in Louisiana. In comparison, the great San Francisco earthquake of 1906 was weaker at 7.8.

Yet the pipeline withstood the powerful quake as designed, damaged but not ruptured, according to the Nov. 10, 2002, Los Angeles Times. If anything, last week's powerful earthquake shows that the pipeline could have withstood more, the pipelines seismic design coordinator said. The New York Times article said that after an aerial survey today, pipeline officials said they found no leaks in the structure.

Gale Norton summarized the effects: The Alaska pipeline was 60 miles from the quake's epicenter. It shook back and forth, some supporting struts broke. But the pipeline held. It did not crack. Not a drop of oil was spilled. No one was injured. The safety systems put in place worked to perfection. The predicted design flaws that supposedly couldn't be overcome by engineering ingenuity weren't mentioned after the earthquake occurred.

Misplaced Effort: Less than five months after the announcement of the oil discovery and proposed pipeline, members of the Sierra Club complained that they were invited to only two superficial meetings where they were told nothing significant, according to The New York Times on July 5, 1969. The Sierra Club and their fellow environmentalists from the Wilderness Society, Friends of the Earth, and Environmental Defense Fund Inc. delayed pipeline progress with lawsuits. The Feb. 13, 1973, New York Times said the delay in construction is the best the oil companies can expect, while the possibility grows ever livelier that after years of misplaced effort the Alaska pipeline will join such forgotten and costly fantasies as the South Sea Bubble.

Reality: That misplaced effort has pumped 15 billion barrels of oil into the U.S. economy. Adrian Herrera of Arctic Power, an Alaska-based group that advocates oil drilling in the Alaska National Wildlife Refuge, said the effects of the pipeline have been huge. The benefit is both economic and social. Infrastructure that was built in conjunction with the pipeline has a trickle-down effect that has helped all businesses. Nationwide the effect has been quite profound, he continued. Not just a direct benefit, there are indirect benefits too. Jobs supporting the pipeline have been spread across the nation, as have the advantages from having more oil available.

Pipeline Breaking: On May 6, 1970, The New York Times said that the head of the Naval Arctic Research Laboratory warned that the proposed trans-Alaska oil pipeline might break and wreak great damage to the environment.

Reality: Despite leaks in the past, the pipeline has improved and is leaking less. The United States has the most stringent environmental controls on oil. Any spill of more than a teaspoon is reported. The whole pipeline is scanned every day from the ground or helicopters for leaks. Despite being three decades old, the pipeline is more modern than many others around the world.

All Hell to Break Loose: The New York Times on Nov 10, 1974, quoted an internationally known professor on Arctic soils from Rutgers University. He predicted all hell will break loose on Alaska's north slope within five years after hot oil starts flowing through the trans-Alaska pipeline. He then compared the spread of damage to the permafrost to a cancer that takes five years.

Reality: Of the 800-mile pipeline, 420 miles are above ground to avoid the permafrost. When above ground, it has a 2-inch heat pipe containing pure ammonia. When the air is cooler than the ground, the ammonia vaporizes and draws the heat from the earth. The ammonia then condenses on the pipe, starting the process again.

Major Oil Incidents Not Caused by Pipeline

The pipeline has not been without accidents but the biggest ones did not involve pipeline malfunctions. On Feb. 15, 1978, there was a leak of 16,000 barrels. There are some indications that it was sabotage. You have to suspect foul play, said Morris Turner of the Alaska Pipeline Office, according to The Washington Post on Feb. 16, 1978. No one was ever charged in that incident. On Oct. 4, 2001, Daniel Carson Lewis, who had been drinking, shot the pipeline and caused a leak of more than 6,000 barrels of oil. The Los Angeles Times on Oct. 21, 2001, quoted a state policeman as saying, "Alcohol and a guy with a gun, nothing deeper than that."

The largest oil-related incident in Alaska since the pipeline was built was the Exxon Valdez incident, not a pipeline failure, but a ship crashing because of human error. On March 24, 1989, a ship hit a reef and spilled more than more than 11 million gallons of crude oil into Prince William Sound. The ships captain, Joe Hazelwood, had been drinking before the ship left, which was illegal. But the time of the ship's departure changed, and had it not, then he wouldn't have broken the law. Hazelwood also left the deck to do other work, leaving the ship with an under-qualified sailor, a breech of company policy.

Exxon's experience with seamen drinking, the union, and the law previous to the Valdez incident.

Randall Fris worked as a seaman on an Exxon Shipping Co. oil tanker for eight years without incident. One night, he boarded the ship for duty while intoxicated, in violation of company policy. This policy also allowed Exxon to discharge employees who were intoxicated and thus unfit for work. Exxon discharged Fris. Under a contract with Fris's union, the discharge was submitted to arbitration. The arbitrators ordered Exxon to reinstate Fris on an oil tanker. Exxon filed a suit against the union, challenging the award as contrary to public policy, which opposes having intoxicated persons operate seagoing vessels

Can a court set aside an arbitration award on the ground that the award violates public policy?

Not only can they, they must.

In this case, the reinstatement would be not only contrary to public policy, but actually illegal. When it turned out that Joe Hazelwood had been convicted of

DUI while on vacation, the USCG amended their rules on alcohol related offenses, **after the grounding of the Exxon Valdez.**

Any person subject to USCG regulation or licensing who commits an alcohol related offense whether on or off duty, or while currently assigned to a vessel or not, is banned for life from working on a US flagged merchant vessel. (There are time periods after which a banned individual may ask to have the ban lifted, but they are long)

Boarding a tanker while intoxicated is a crime. Were Exxon to allow him to work on one of their tankers, they would be committing a crime - and "An arbitrator told us to reinstate him" would NOT be a defense.

It is important to know that this was only true after the Valdez accident and after the law was changed because of the accident. One more case of government locking the barn door after the horse has been stolen.

Questions: How did Hazelwood get aboard while drunk? Are sailors required to take a sobriety test before coming on board? Who checks on this anyway? What prevents a sailor from becoming drunk while on board? Are there checks, procedures, and/or laws to prevent this?

There have been many instances where firing of a seaman for an infraction of the rules and law was reversed with penalties to the company when a union arbitrated a dispute or sued. A number of these cases resulted in disasters subsequently blamed on "the company" who then had to pay for the damage. It is quite obvious that the main reason a company is blamed and punished for actions of and/or by an employee that are strictly forbidden by reasonable company policies, is that the company has large financial resources and the employee does not. Lawyers will always go after the *deep pockets* in these cases. The general public loves to see those evil capitalists foot the bill and the media will add as much fuel to that fire as their fertile imaginations can dream up. I call it their *Simon Lagree* complex.

Aftermath: While many animals were killed and the environment was damaged, it was not nearly as devastating as the "twenty year death zone" and "end of salmon fishing" predicted for Prince William Sound by many overreacting environmental activists. The area has since bounced back giving evidence of the resilience of life on our planet. The 2005 salmon run was so large that millions of fish were left to die and rot in hatchery areas. Exxon has paid out $3.5 billion in relation to the oil spill. Alyeska, a consortium of oil companies of which Exxon is a part, spends around $60 million a year on oil spill prevention in Prince William Sound.

The real lesson all of this provides is the revelation of the ridiculous power of unions to abrogate even the most necessary safety regulations while often transferring the blame from drunk and incompetent members to the companies, even to the point of reinstating guilty members after they have broken company rules and the law with terrible consequences. There are several reasons for this reality. One is the fact that suing a person as compared to a corporation with deep pockets offers no promise of wealth to attorneys. Another is that unions and their member have little or no accountability for their actions under the law, even criminal actions. They cannot be sued like corporations or individuals. Such is the political power unions have with their partners in crime in the liberal camp. If the unions could be held even partially legally responsible for their actions or complicity in these cases, things would be a lot different and certainly safer for everyone. Environmental disasters like the Exxon Valdez spill would be much less likely to happen.

NPOs - Problems - Research - Solutions - Solvers - Users

How They Relate to Global Warming/Climate Change

PROBLEMS AND THEIR FALLOUT
The Global Warming Promoters.

I start with these word definitions so the reader will better understand the basis for my conclusions.

Non Profit Organization: abbreviated **NPO**, also known as a not-for-profit organization, is an organization that does not distribute its surplus funds *(actually profits, that nasty word)* to owners or shareholders, but instead supposedly uses them to help pursue its goals. It can, and usually does also own and operate property including buildings, equipment and vehicles. Employees, including owners and managers, of NPOs receive salaries and other compensation, often quite substantial. Actually, non profit status provides the opportunity for much financial mischief by the owners and/or operators of such entities.

Consider this: Many small, closely-held corporations, often family businesses, operate as an NPO without NPO status, an "S" corporation. All the owners need do is pay themselves in salary whatever profit the corporation made during the year so the corporation breaks even or only shows a tiny profit. This is perfectly legal and is a practice of many of these corporations. Of course, it does not happen with any publicly held corporations, at least not intentionally. NPOs operate exactly the same way. Any profits they might show can be used to further the purpose of the organization. In actuality, they are often distributed as pay or bonuses to the owners or operators. There are some rules they must follow, but basically they become corporate tax-free operations, exactly like the small, closely-held private corporations. Of course, both of them pay taxes in the form of personal income taxes, so they are not true tax free operations. Unlike normal corporations, true

NPOs can claim to be profit-free and get away from that word, *profit*, so despised by liberals, the media, and most of the public. Their owners can then smile all the way to the bank without fear of being branded as **greedy capitalists**.

problem: a question raised for inquiry, consideration, or solution - an intricate, unsettled question - a source of perplexity, distress, or vexation - a difficulty in understanding or accepting.

research: careful or diligent search - studious inquiry or examination; *esp*: investigation or experimentation aimed at the discovery and interpretation of facts, revision of accepted theories or physical laws in the light of new facts, or practical application of such new or revised theories or laws - the collecting of information about a particular subject.

solution: an action or process of solving a problem - an answer to a problem - a bringing to an end or into a state of discontinuity.

solver: one who solves a problem or brings it to a successful solution.

user: one who uses problems and avoids solutions primarily for their own personal benefit.

Problems, and the methods used to deal with them, are as varied as are the thoughts and ideas of individual humans. Someone once said that if you could adequately define a problem it would almost certainly lead to a conclusion and thus to a solution. The very word problem, however, implies the unknown.

Everyone deals with problems, usually on a daily basis. Problems can be divided into two types: those that have solutions and those that do not. Knowing the difference, and in which of the two categories a problem belongs is a problem in itself. For less than obvious reasons different people and different groups deal quite differently with problems. While the specific approach is not the purview of this writing, there are generally two very different classes of how to deal with any problem.

Some view a problem as something to be solved so that it is no longer a problem, while others view the same problem as something to be maintained, nurtured and kept alive without solution for their own purpose. To the latter group, any solution is to be avoided at all costs as that would end the usefulness of the problem to their purposes.

In general, engineers, contractors, business people, professionals in many fields, those in industry, and even individual entrepreneurs—mostly private sector people—deal with problems using the first method. They work hard at finding a viable and usually profitable solution, and move onto other things once it is solved. The result is all of the marvelous technologies and the organizations that developed and manufactured them, and that have so improved and extended our lives. To these people, a problem is something to be solved. Their attitudes about solutions to problems created the industrial and technological revolution that has resulted in everything from ipads to space missions to providing safe foods and excellent health care for billions of humans.

Members of the other group, who do not want solutions, want every problem to go no forever without solution. Why? Consider the politician. Every problem is viewed as a weapon with which to bludgeon political opponents, or obtain money. Should such a problem be solved it would remove a valuable tool from his or her political arsenal. That's why politicians are problem *users* and not problem *solvers*. Politicians love unsolvable problems, especially those they can use to impose taxes. Global warming is a classic example that is trumpeted in the main stream media on a daily basis. It has become a huge financial boon for some and a ticket to power for others. This is why many groups, of government bureaucrats in particular, are organized to solve problems that are extremely difficult to solve. Global

warming is currently the best and biggest prospect as a problem known to be impossible to solve. There are many of these problems that are the objects of effort by entire departments of governments and especially the UN.

Example: Government bureaucrats comprise a large group who do not want solutions to the problems they are charged with solving. Consider this situation: a government agency has a small department that is dealing with a specific problem, a common situation. As manager, part of your compensation is determined by how big a budget you can get for your department. One of the workers in the department comes to you with the perfect solution to the problem—problem solved, no further need for the department to exist. You have a choice. The obvious choice is to implement the solution, dissolve the department, and look for another job. Is this what happens? Not on this planet. The more likely scenario is that you thank the employee and then transfer him to another department as far from yours as possible. You then destroy all records of the solution and go back to working on the problem, business as usual.

This brings us to another group that will avoid any solution at all costs. These are government and university research organizations, particularly those that vie for grants mostly provided by the US government. As long as the problem has legs it will be used to obtain grant money to fund the group's research. Should the problem the group is dealing with be solved, there is no longer a need for the group to exist so it will be disbanded, the grant will be cancelled, and the group members will lose their jobs, right? Well . . . not always, especially if it is a government or government-sponsored research organization. These groups are like the mythical Hydra, chop off one head and two more appear. An in-depth examination of government records will unearth many groups and committees whose usefulness ceased long ago, some as far back as horse and buggy days. These groups continue to have their budgets increased every year thanks to a Congress enamored of base line budgeting and pork barrel projects for their districts or states. Every once in a while someone discovers one of these dinosaurs and blows a whistle. If it would happen to be a Congressman, whatever he discloses had better not be of benefit to his district or his opponent will use it to destroy him in the next election.

Even in universities it is difficult to get a research group disbanded once they have successfully run the grant committee gauntlet. Politics plays a huge part even there. I had an email exchange with a friend, a member of one of my Internet writers critique groups, who is a professor at a European university. For a short time he was a member of one research group. He left in disgust after discovering the purpose of the group had long ago ceased to exist and that their sole purpose seemed to be in writing grant requests so they would not be disbanded. Of course, it really wasn't quite that simple.

My friend's expertise and main pursuit was research on the effects of increased atmospheric carbon dioxide on European plants and specifically crop plants. He kept getting refused new government grants until he changed the name of his research. He changed the title of his grant request from what previously was, *The Beneficial Effects of Increased Atmospheric Carbon-dioxide on European Crop Plants*, to the infinitely more politically correct, *The Deleterious Effects of Increased Atmospheric Carbon-dioxide on European Crop Plants*. He said he also changed a few of the sub headings in a similar fashion and his grant came through. His research had not changed, his conclusions had not changed and with a few insignificant revisions to parts of his conclusions, he published his results, some of which I included in my book, *Energy, Convenient Solutions.*

I could list other example, but I'm sure you can see the basic differences between the two groups. Those who do not want problems solved are probably in the minority in universities, but I will wager they represent a majority in government research organizations and in many of the so called *think tanks* that have sprouted up all over. I have two examples of why I believe this to be true

The Postal Service: At least thirty years ago, I was flying to Chicago from Des Moines Iowa. I struck up a conversation with the man seated next to me who happened to be the Postmaster of Chicago. He was a man in his mid sixties who informed me he was about to retire. Then he told my why he was glad he was retiring and getting out of the Post office.

It seems he and several of his subordinates had developed a new training program to help low income, entry level employees, primarily minorities, get training so they could work for the Post Office. This was done at the request of the head of the post office in Washington. Their pilot program was started with twenty applicants that were not required to take any tests. In fact, some of them could not even read. The program was a four-week intensive training given by Post Office employees. The trainees were paid the same as entry level jobs during their training. At the end of the program they were each given the standard Post Office test for the jobs they were to do. Eleven passed and went to work. At this time eleven new applicants were added to the nine that did not pass and the entire group went through the training process.

Again they were given the standard employment test. Seven of the new people and four of the ones who went through twice, passed, and went to work. Once more there were nine left who would retake the training. The program called for applicants to be let go at the end of the third training period if they did not pass the tests. This time only four of the new people passed along with two of the second timers. That meant that only six new applicants could be accepted. At the end of the third training period, two of the new applicants passed along with one of the repeaters. Those who had taken the third training and still couldn't pass were released. At the end of the next program five more were released. At this point someone, the Postal Workers Union or the NAACP, or the ACLU, I don't remember who, but one of those filed a suit to force the post office to keep all trainees in the program until they were able to pass the test.

The results were obvious. The program was soon filled with individuals who would never be able to pass the test and thus became permanent non working employees. At the time of our flight, the program had been disbanded. The Postmaster was terribly disappointed saying, "Our program really worked and was providing jobs for a few who were otherwise considered unemployable. We were about to expand it when those lousy do-gooders got in and messed it up. Now we are again without a training program for these special entry-level people. They are the ones who ultimately suffered."

As far as I know those twenty non working employees could still be there, doing nothing and drawing a government paycheck. Remember this was more than thirty years ago so my numbers may not be on the money and I'm sure his words were different from my quotes, but the story itself is true.

Infinitely Politically Correct Anthropogenic Global Warming:

This foolishness has achieved such power that nearly all of the news media, politicians, academics, intellectual elitists, and virtually all of the misinformed public treat it as a scientifically proven fact. This is nonsense. It is nothing but *consensus* science driven and tilted by politics and the lure of money—lots of money. The really sad part of this whole thing is the polarization it has brought about in the otherwise usually objective scientific community and the valuable effort diverted from countering some truly dangerous menaces, the real problems facing us and their real solutions. If you question any aspect of the global warming mantra you will be ridiculed and

called names (*as I have been*) and ostracized. *(They can't do that to me, but can to others.)*

Witness what happened to Dr. Judith Curry, head of the School of Atmospheric Sciences of Georgia Tech. Just because she will not condemn 100% of those who question the efficacy of the IPCC, the Intergovernmental Panel on Climate Change, she has been the subject of many verbal attacks. To my great admiration, she has not backed down. She has accused the IPCC of, corruption and says, "I'm not going to spout off and endorse the IPCC because I don't have confidence in the process." This was before the Climategate email revelations of doctored computer simulations.

She has been jeered, insulted and otherwise badly treated, because she doesn't knuckle under to the pressures of the PC church of global warming. Incidently, she's not a denier, just a questioning skeptic, as am I, who would like a whole lot more proof. I looked into what she is asking and her questions are virtually the same ones I have been asking. Where is the hard science, the physics and chemistry, that proves global warming from carbon-dioxide is as real as proponents of the *theory* say it is? What about the many other factors that effect global temperatures? How about Svensmark's theory? (See Pages 16-26)

I am reminded of one of my favorite quotes from Peter Abelard, "By doubting we are led to inquire. By inquiring we learn the truth." Incidentally, Abelard's words are a quite good description of the *scientific method*. Apparently the global warming crowd will tolerate no doubting or inquiring from anyone, even highly qualified scientists. I do not believe that is even a remotely scientific attitude. Good science welcomes doubters and questioners. As a matter of fact, the entire basis of science, the *scientific method*, is based on repeated and thorough questioning. The idea that *consensus science* (a group opinion) is superior to *hard science* (math, physics, chemistry) is ludicrous. It goes against the grain of all true scientific reasoning, and in truth, **the opposite is always true. Hard science always trumps consensus science.** This does not mean that *consensus* science is wrong or is not a valuable tool. It merely means that it is a consensus opinion of a group of scientists, a group that could even be a minority of scientists. Sometimes it is all we have when hard scientific evidence is lacking. (Not the case with global warming.)

This is my interpretation from a pointedly unflattering and somewhat misleading article about Dr. Curry in the November 2010 issue of Scientific American. The title, *Climate Heretic*, is insulting. There is a full page photo of Dr. Curry opposite the title page that is also less than flattering. The only reference to her well-earned title is the following on the first paragraph. "For most of her career, Curry, who heads the School of Earth and Atmospheric Sciences at the Georgia Institute of Technology, has been known for her work on hurricanes, Arctic ice dynamics, and other climate related topics. But over the past years or so she has become better known for something that annoys, even infuriates many of her scientific colleagues, probably because it might threaten their grants. (*She refuses to go along with the crowd like a sheep.*) Curry has been engaged actively with the climate change skeptic community, largely by participating on outsider blogs such as *Climate Audit*. *The Air Vent* and *The Blackboard*. Along the way, she has come to question the science, no matter how well established it is." *(The consensus science promoted by many of those at the global warming money trough.)*

In typical misleading fashion, *Scientific American* printed a version of the *hockey stick* graph of global temperatures (see first graph, on page 13) showing best guess temperatures from 1000 CE to the present. Since the **Little Ice Age** began around 900 CE, well before the dates on the graph, this version, commonly used to illustrate how temperatures have risen in recent years, gives an extremely erroneous picture. If you compare it with a similar graph starting say 10,000 BC, a very different picture appears. It is quite plain from the expanded data that current global temperatures are considerably lower than those during the **Medieval Warm Period** from 200 CE to 900 CE and even warmer at several times since the last ice age. Why is it that global warmers never refer to this data and will certainly not show these graphs?

The author of the article is one I consider to be a dedicated member of the fundamentalist church of global warming, a political hack writer. His name is Michael D. Lemonick and he was a longtime far-left science writer for Time Magazine. He now writes for Climate Central, a non profit, non partisan climate change think tank. (I about choked on that oxymoronic description, *non partisan*. I have come to recognize the label *non partisan* as an indication of extreme partisanship of any person, organization, or proposal. Like many other politically correct labels, it means precisely the opposite of what it says. This type of mislabeling is a favorite tactic of liberals. Another example is *The Affordable Care act*.) Climate Central is one of the hundreds of usually non profit organizations that have sprung up to feed on the leavings from the global warming fantasy promoters. The Internet is loaded with them, all soliciting donations for their noble purpose. They may be non profit, but I'll wager their principals receive a hefty paycheck along with many perks. Many NPOs have highly paid executives who fly around in private jets. (Like Nancy Pelosi whose NPO was Congress) NPOs can offer all the perks of any profit making corporation for their owners and employees. The only difference is they don't pay their owners in dividends, they simply pay them in salary and benefits.

OK, so I rambled about. I wanted to share some realities from the wonderful world of liberalism.

The following is a response to the article on Dr. Curry from one Climatologist made on 10-23-2012

This article completely neglects to mention the enormous amounts of grant money being shoveled into **climate studies**. $Billions every year are handed out by the federal government, with much more payola coming from shadowy, politically oriented NGOs that are often at odds with honest science.

Big money corrupts, as can be seen throughout the **Climategate** emails, where journals are threatened and blackballed, and journalists and FOI officers are corrupted, and professional careers are ruined, simply for not toeing the alarmist line. The mainstream climate clique has both front feet in the public grant trough, and it brazenly shoulders aside scientific skeptics (the only honest kind of scientists, according to the scientific method).

Dr. Curry has taken a brave stand, breaking ranks with the current orthodoxy. She is a finger to the wind, indicating a sea change in the public's growing awareness of the fact that there is zero credible evidence showing that the rise in CO_2 has been harmful while there is solid, testable, empirical evidence showing that the rise in CO_2 has been beneficial, such as increasing agricultural production in a world that needs more food.

The IPCC has become entirely self-serving since AR-1. It is now much more interested in protecting its grant gravy train than in allowing skeptical scientists to be a part of the process. It took knowledgeable outsiders to debunk Michael Mann's *hockey stick* chart; the iconic poster of the IPCC.

In retrospect, the scientific establishment should have promptly sounded the alarm when it was claimed in MBH98-99 that the planet's temperature was essentially unchanging over many centuries. Instead, the Mann et al. attempt to erase the MWP and the LIA was unquestioningly accepted, at least publicly, due to the immense flow of grant money at stake. Further, the IPCC still continues to avoid the scientific method, instead protecting its catastrophic AGW hypothesis from any and all attacks by skeptical scientists. Since when has it become the duty of scientists to falsify hypotheses?

But the cracks in the defenses of the climate alarmists are widening. Taxpayers are disgusted with the unaccountable hand over fist money grabbing by a completely unaccountable UN/IPCC. As the public becomes more aware of how the system is being gamed at their expense, push back is increasing. And it will continue to escalate.

End of the climatologist's response

You see, I'm not the only one. Increasing numbers of people are asking all kinds of probing and even embarrassing questions of climate alarmists. A lot of this was triggered by the revelations of doctored computer simulations by the famous hackers who found much evidence of mischief in University computer files. My friend, Levelhead, who is a professor at East Anglia University, told me the situation was even worse than reported by the media.

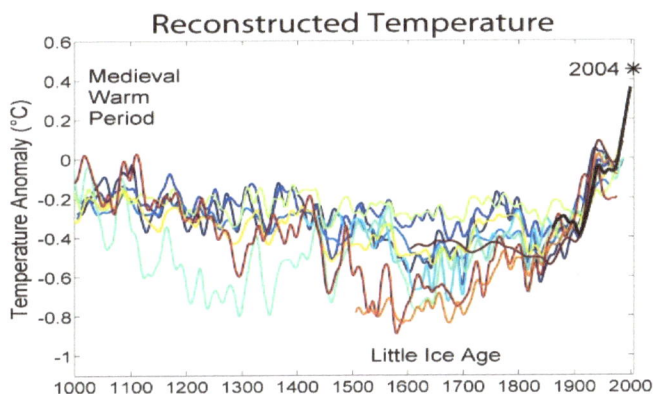

Here are some graphs the reader may find interesting.

Above is the famous IPCC *hockey stick* graph shown in an accurate scale. Compare it with the same period shown on the graph on the next page. Notice only one line, the black one, goes up sharply at the end. This line represents tree ring growth for the last 150 years that the IPCC uses erroneously to show temperatures. It is one of the eight types of temperature estimates shown, and is highly unreliable. A far more logical explanation of expanding tree ring growth in recent times is that the increased amount of carbon dioxide in the atmosphere is the direct cause of the increased growth.

It is interesting to note that the high point in the first graph (the black line) is based on tree ring growth data. It is very difficult to correlate the various other well-documented representations of temperatures calculated from various indirect measurements in atmospheric CO_2 which has been found to greatly increase plant growth including trees. This holds true almost without regard for temperature, so tree ring data is totally useless as a measure of ambient temperatures unless the data is corrected for changes in atmospheric CO_2. You will also note that the black line is only shown for the last 150 years. If it were shown for the full thousand years, it's irregular meanderings would prove it to be totally meaningless as an indicator of temperature. A careful look at the graph reveals that several of the colored lines disappear around 1960 and only the black line for tree ring measurements is shown to 2000. That is because almost all the other lines would show a decided downward trend of temperatures before 2000.

To see how the graphs were manipulated by **AGW proponents, view the graphs on page 33.** These show what the real numbers indicate, and how they were manipulated to paint a false picture of supposed AGW. (Anthropogenic Global Warming)

This graph is a more telling graph of temperatures since the end of the last ice age. As you can see, the tree ring data (red line) wanders a path very different from most of the others. It is the only one that goes up sharply at the zero point. Most of the indicators are well below the zero line and one (light blue) is farther below the median at zero than at any time since the ice age. It is obvious that most of the preceding 8,000 years have been warmer than the present and that 400 years ago we reached lower temperatures than the earth has seen since the last ice age. It is interesting to note that at the time Stonehenge and the Pyramids were built the earth was a lot warmer and probably a lot wetter than it is today. Most of that warmth flowed into the north and south temperate zones while the tropics remained much the same as it is today. In fact, the tropics probably stayed relatively warm even at the ice age minimum temperatures.

This next graph is the IPCC temperature graph before Michael Mann published his *hockey stick* graph and report that completely ignored the ***Medieval Warm Period***, the ***Little Ice Age***, and exaggerated the effects of increased atmospheric CO_2.

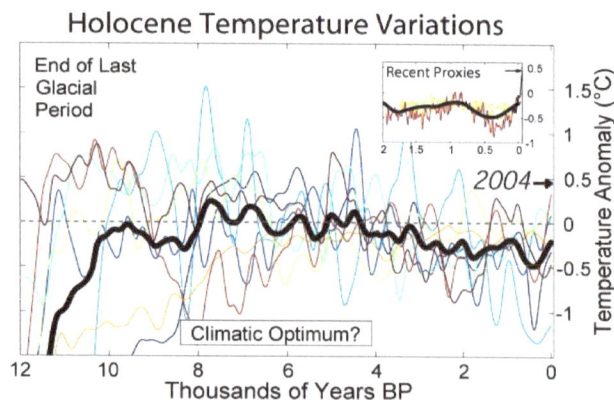

For more information on the hockey stick graph goto: http://www.john-daly.com/hockey/hockey.htm

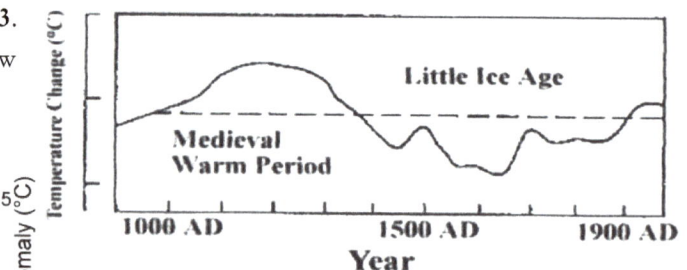

This next graph is from data collected during the Vostok ice core study. It plots temperatures and Carbon dioxide content of the atmosphere back through the last four ice ages. As you can see, an earth much colder than the present has been the rule for many millennia. Clearly, cold periods are quite long while temperate periods like the present are relatively short. If past cycles repeat as it looks like they will, we are about due to drop into the deep freeze in the near future. If it has any effect, man's addition of CO_2 to the atmosphere might temper the next ice age. This graph is a far better indicator of where we will be going in the future than the others.

It is fairly obvious and thus quite definitive that the CO_2 variation (the green line) **follows** the average temperature (the blue line) rather than the other way around. This significant indication shown clearly on the graph indicates that increases and decreases in CO_2 content **follow** increases and decreases in temperature and do not precede them. This is a positive and convincing indication that changes in the amount of CO_2 are the **result** of temperature changes, not the cause.

Global Warming - ACS - excerpt - 12-29-07 +new
http://glowarmacs.blogspot.com

Global Warming and the Gulf Stream Facts & Facts
http://hjgulfstream.blogspot.com

Global Warming and Earth Hour
http://hjglobalwarming.blogspot.com

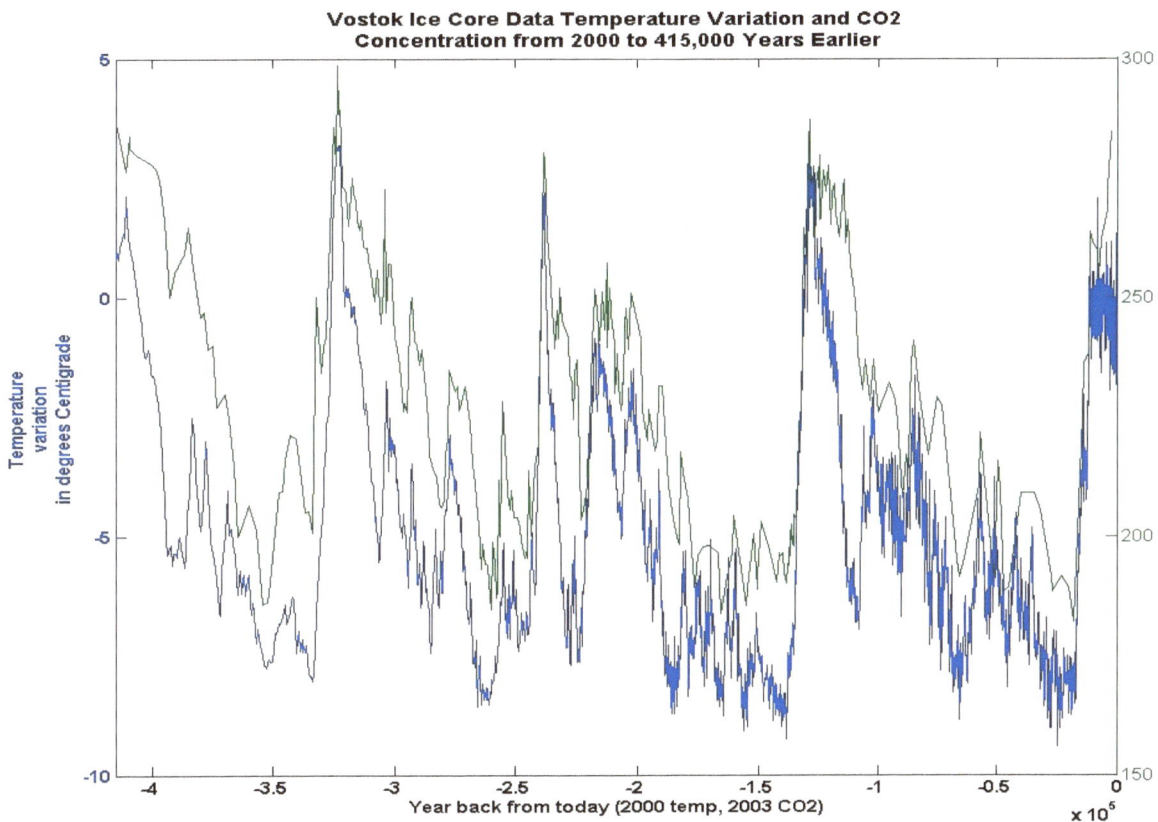

Vostok Ice Core Data Temperature Variation and CO2 Concentration from 2000 to 415,000 Years Earlier

A stellar revision of Global Warming Realities

Climate Change: News and Comments and The Svensmark Hypothesis

The Photo - The Pleiades - The Seven Sisters

Visible to the naked eye as the Seven Sisters, the Pleiades are the most famous of many surviving clusters of stars that formed together at the same time. The Pleiades were born during the time of the dinosaurs, and the most massive of the siblings would have exploded over a period of 40 million years. Their supernova remnants generated cosmic rays. From the catalogue of known star clusters, Henrik Svensmark has calculated the variation in cosmic rays over the past 500 million years, without needing to know the precise shape of the Milky Way Galaxy. Armed with that astronomical history, he digs deep into the histories of the climate and of life on Earth.

(Image ESA/NASA/Hubble)

Nigel Calder

The following are quotes from Svensmark's Cosmic Jackpot and other comments by Nigel Calder. Nigel Calder is the author of a marvelous science book, *The Magic Universe*. Note the British spelling in his article.

The Royal Astronomical Society in London published (online) Henrik Svensmark's paper entitled **"Evidence of nearby supernovae affecting life on Earth."** After years of effort Svensmark shows how the variable frequency of stellar explosions not far from our planet has ruled over the changing fortunes of living things throughout the past half billion years. Appearing in Monthly Notices of the Royal Astronomical Society, It's a giant of a paper, with 22 figures, 30 equations and about 15,000 words.

See the RAS press release:

http://www.ras.org.uk/news-and-press/219-news-2012/2117-did-exploding-stars-help-life-on-earth-to-thrive

By taking me back to when I reported the victory of the pioneers of plate tectonics in their battle against the most eminent geophysicists of the day, it makes me feel 40 years younger. Shredding the textbooks, Tuzo Wilson, Dan McKenzie and Jason Morgan merrily explained earthquakes, volcanoes, mountain-building, and even the varying depth of the ocean, simply by the drift of fragments of the lithosphere in various directions around the globe.

In Svensmark's new paper an equally concise theory that cosmic rays from exploded stars cool the world by increasing the cloud cover, leads to amazing explanations, not least for why evolution sometimes was rampant and sometimes faltered. In both senses of the word, this is a stellar revision of the story of life.

Here are the main results:

The long-term diversity of life in the sea depends on the sea-level set by plate tectonics and the local supernova rate set by the astrophysics, and on virtually nothing else.

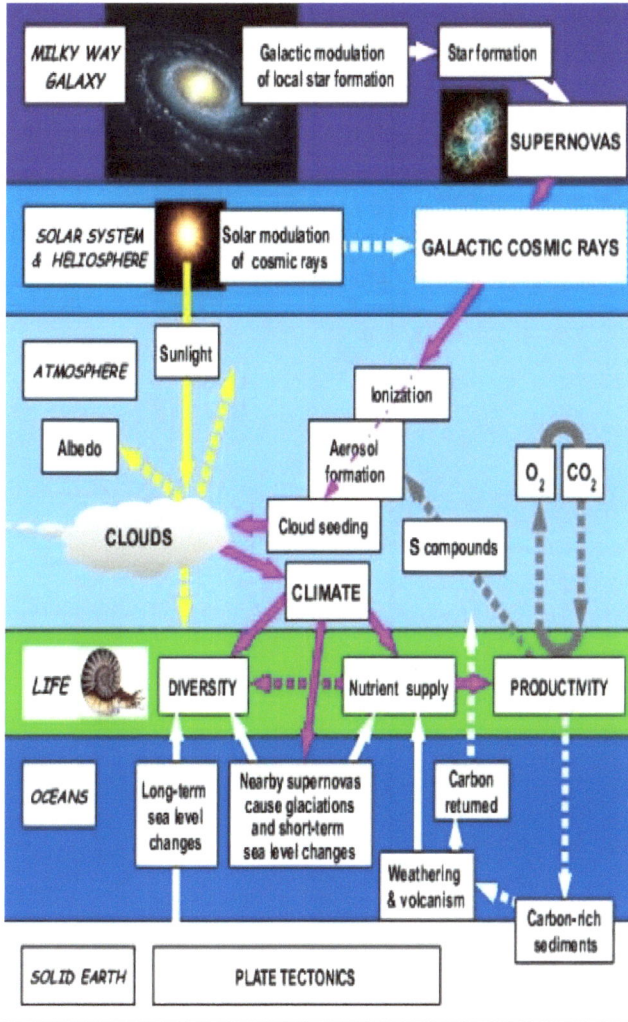

© Nigel Calder 2012. Credits: Milky Way impression NASA; Crab Nebula NASA-ESA/Hubble; Sun image NOAA; Ammonite impression San Diego Natural History Museum.

The long-term primary productivity of life in the sea – the net growth of photosynthetic microbes – depends on the supernova rate, and on virtually nothing else.

Exceptionally close supernovae account for short-lived falls in sea-level during the past 500 million years, long-known to geophysicists but never convincingly explained.

As the geological and astronomical records converge, the match between climate and supernova rates gets better and better, with high rates bringing icy times.

Presented with due caution as well as with consideration for the feelings of experts in several fields of research, a story unfolds in which everything meshes like well-made clockwork. Anyone who wishes to pooh-pooh any piece of it by saying "correlation is not necessarily causality" should offer some other mega-theory that says why several mutually supportive coincidences arise between events in our galactic neighborhood and living conditions on the Earth.

NOTE: *correlation of increased CO_2 and global warming are several orders weaker than Svensmark's graphical data.*

An amusing point is that Svensmark stands the currently popular carbon dioxide story on its head. Some geoscientists want to blame the drastic alternations of hot and icy conditions during the past 500 million years on increases and decreases in carbon dioxide, which they explain in intricate ways. For Svensmark, the changes driven by the stars govern the amount of carbon dioxide in the air. **Climate and life control the amount of CO_2 in the atmosphere, not the other way around.** The catastrophists have it backwards.

By implication, supernovae also determine the amount of oxygen available for animals like you and me to breathe. So the inherently simple cosmic-ray/cloud hypothesis now has far-reaching consequences, which I've tried to sum up in this diagram.

Cosmic rays in action. The main findings in the new Svensmark paper concern the uppermost stellar band, the green band of living things and, on the right, atmospheric chemistry. Although solar modulation of galactic cosmic rays is important to us on short time scales, its effects are smaller and briefer than the major long-term changes controlled by the rate of formation of big stars in our vicinity, and their self-destruction as supernovae. Although copyrighted, this figure may be reproduced with due acknowledgment in the context of Henrik Svensmark's work.

By way of explanation

The text of "Evidence of nearby supernovae affecting life on Earth" is available via

ftp://ftp2.space.dtu.dk/pub/Svensmark/MNRAS _Svensmark2012.pdf.

The paper is highly technical, as befits a professional journal, so to non-expert eyes even the illustrations may be a little puzzling. So I've enlisted the aid of Liz Calder to explain the way one of the most striking graphs, Svensmark's Figure 20, was put together. That graph shows how, over the past 440 million years, the changing rates of supernova explosions relatively close to the Earth have strongly influenced the biodiversity of marine invertebrate animals, from trilobites of ancient times to lobsters of today. Svensmark's published caption ends: "Evidently marine biodiversity is largely explained by a combination of sea-level and astrophysical activity." To follow his argument, you need to see how Figure 19 draws on information in Figure 20. That tells of the total diversity of the sea creatures in the fossil record, fluctuating between times of rapid evolution and times of recession. The count is by genera which are groups of similar animals. Here it's shown freehand by Liz in Sketch A. Sketch B is from another part of Figure 20, telling how the long-term global sea-level changed during the same period.

The broad correspondence isn't surprising because high sea-levels flood continental margins and give the marine invertebrates more extensive and varied habitats. But it obviously isn't the whole story. For a start, there's a conspicuous spike in diversity about 270 million years ago that contradicts the declining sea-level. Svensmark knew that there was a strong peak in the supernova rate around that time. So he looked to see what would happen to the wiggles over the whole 440 million years if he normalized the biodiversity to remove the influence of sea-level. That

simple operation is shown in Sketch C, where the 270-million-year spike becomes broader and taller. Sketch D shows Svensmark's reckoning of the changing rates of nearby supernovae during the same period. Let me stress that these are all freehand sketches to explain the operations, not to convey the data. In the published paper, the graphs as in C and D are drawn precisely and superimposed for comparison.

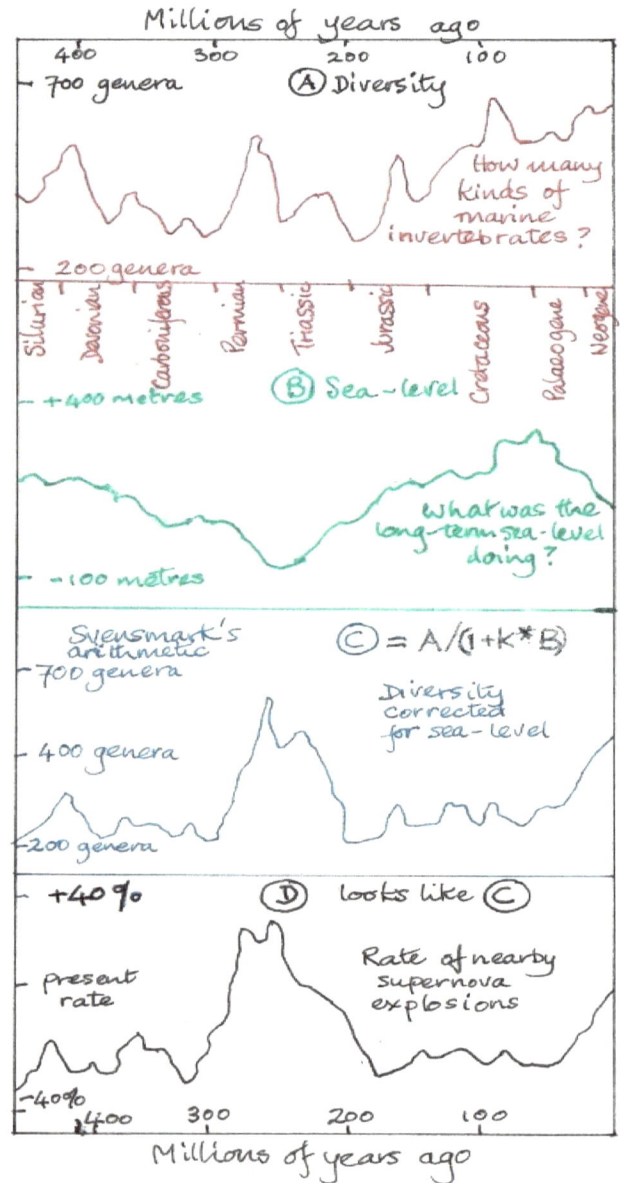

Figure 19

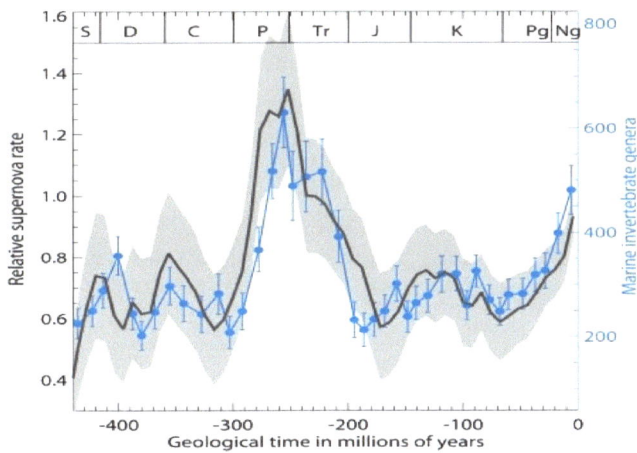

Above is Svensmark's Figure 20, with axes relabeled with simpler words for the RAS press release. Biodiversity (the normalized marine invertebrate genera count) is in blue, with vertical bars indicating possible errors. The supernova rates are in black.

There are many fascinating particulars that I might use to illustrate the significance of Svensmark's findings. To choose the Gorgon's story that follows is not entirely arbitrary, because this brings in another of those top results, about supernovae and bio-productivity.

The great dying at the end of the Permian

Out of breath, poor gorgon? Gasping for some supernovae? Named after scary creatures of Greek myth, the Gorgonopsia of the Late Permian Period included this fossil species Sauroctonus progressus, three metres long. Like many of its therapsid cousins, near relatives of our own ancestors, it died out during the Permo-Triassic Event.

Source:
http://en.wikipedia.org/wiki/Gorgonopsia

Luckiest among our ancestors was a mammal-like reptile, or therapsid, that scraped through the Permo-Triassic event, the worst catastrophe in the history of animal life. The climax was 251 million years ago at the end of the Permian Period. Nearly all animal species in the sea went extinct, along with most on land. The event ended the era of *old life*, the Paleozoic, and ushered in the Mesozoic Era, when our

ancestors would become small mammals trying to keep clear of the dinosaurs. So what put to death our previously flourishing Gorgon-faced cousins of the Late Permian? According to Henrik Svensmark, the Galaxy let the reptiles down.

The Poor Gorgon

Forget old suggestions (by myself included) that the impact of a comet or asteroid was to blame, like the one that did for the dinosaurs at the end of the Mesozoic. The greatest dying was less sudden than that. Similarly the impressive evidence for an eruption 250 million years ago—a flood basalt event that smothered Siberia with noxious volcanic rocks covering an area half the size of Australia—tells of only a belated regional coup de grâce. It's more to the point that oxygen was in short supply. Geologists speak of a "superanoxic ocean" where there was far more carbon dioxide in the air than there is now.

"Well there you go," some people will say. "We told you CO_2 is bad for you." That, of course, overlooks the fact that the notorious gas keeps us alive. The recently increased CO_2 shares with the plant breeders the credit for feeding the growing human population. Plants and photosynthetic microbes covet CO_2 to grow. So in the late Permian its high concentration was a symptom of a big shortfall in life's productivity, due to few supernovae, ice-free conditions, and a lack of weather to circulate the nutrients. And as photosynthesis is also badly needed to turn H_2O into O_2, the doomed animals were left gasping for oxygen, with little more than half of what we're lucky to breathe today.

When Svensmark comments briefly on the Permo-Triassic Event in his new paper, "Evidence of nearby supernovae affecting life on Earth," he does so in the context of the finding that high rates of nearby supernovae promote life's productivity by chilling the planet, and so improving the circulation of nutrients needed by the photosynthetic organisms.

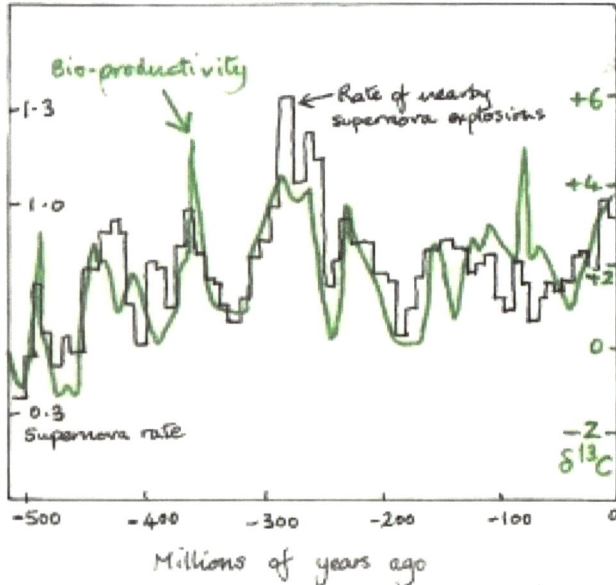

Above is a sketch from Figure 22 in the paper. It is simplified to make it easier to read. Heavy carbon, 13C, is an indicator of how much photosynthesis was going on. Plumb in the middle is a downward pointing green dagger that marks the Permo-Triassic Event. And in the local supernova rate (black curve) Svensmark notes that the Late Permian saw the largest fall in the local supernova rate seen in the past 500 million years. This was when the Solar System had left the hyperactive Norma Arm of the Milky Way Galaxy behind it and entered the quiet space beyond. "Fatal consequences would ensue for marine life," Svensmark writes, "if a rapid warming led to nutrient exhaustion . . . occurring too quickly for species to adapt."

One size doesn't fit all, and a fuller story of Late Permian biodiversity becomes subtler and even more persuasive. About six million years before the culminating mass extinction of 251 million years ago,

a lesser one occurred at the end of the Guadalupian stage. This earlier extinction was linked with a brief resurgence in the supernova rate and a global cooling that interrupted the mid-Permian warming. In contrast with the end of the Permian, bio-productivity was high. The chief victims of this die-off were warm-water creatures including gigantic bivalves and rugose corals.

Why it's tagged as "astrobiology"

So what, you may wonder, is the most life-enhancing supernova rate? Without wanting to sound like Voltaire's Dr. Pangloss, it's probably not very far from the average rate for the past few hundred million years, nor very different from what we have now. Biodiversity and bio-productivity are both generous at present.

Svensmark has commented (not in the paper itself) on a closely related question – where's the best place to live in the Galaxy?

"Too many supernovae can threaten life with extinction. Although they came before the time range of the present paper, very severe episodes called Snowball Earth have been blamed on bursts of rapid star formation. I've tagged the paper as *Astrobiology* because we may be very lucky in our location in the Galaxy. Other regions may be inhospitable for advanced forms of life because of too many supernovae or too few."

Astronomers searching for life elsewhere speak of a Goldilocks Zone in planetary systems. A planet fit for life should be neither too near to nor too far from the parent star. We're there in the Solar System, sure enough. We may also be in a similar Goldilocks Zone of the Milky Way, and other galaxies with too many or too few supernovae may be unfit for life. Add to that the huge planetary collision that created the Earth's disproportionately large Moon and provided the orbital stability and active geology on which life relies, and you may suspect that, astronomically at least, Dr. Pangloss was right — "Everything is for the best in the best of all possible worlds."

Don't fret about the diehards

If this blog has sometimes seemed too cocky about the Svensmark hypothesis, it's because I've known what was in the pipeline, from theories, observations and experiments, long before publication. Since 1996 the hypothesis has brought new successes year by year and has resisted umpteen attempts to falsify it.

New additions at the level of microphysics include a previously unknown reaction of sulphuric acid, as in a recent preprint. On a vastly different scale, Svensmark's present supernova paper gives us better knowledge of the shape of the Milky Way Galaxy.

A mark of a good hypothesis is that it looks better and better as time passes. With the triumph of plate tectonics, diehard opponents were left red faced and blustering. In 1960 you'd not get a job in an American geology department if you believed in continental drift, but by 1970 you'd not get the job if you didn't. That's what a paradigm shift means in practice and it will happen sometime soon with cosmic rays in climate physics.

Plate tectonics was never much of a political issue, except in the Communist bloc. There, the immobility of continents was doctrinally imposed by the Soviet Academy of Sciences. An analogous diehard doctrine in climate physics went global two decades ago, when the Intergovernmental Panel on Climate Change was conceived to insist that natural causes of climate change are minor compared with human impacts.

Don't fret about the diehards. The glory of empirical science is this: no matter how many years, decades, or sometimes centuries it may take, in the end the story will come out right.

Sorry folks, cosmic rays really are in charge

On this blog and others, most comments about my previous post "Yet another trick of cosmic rays" have been friendly. Thank you. But some people still want to dismiss all the meticulous experimental, observational and theoretical work of Henrik Svensmark and his colleagues in the Danish National Space Institute by saying there is simply no link between cosmic rays and the climate.

Having written two books on the subject, and still engaged with it, I could in rebuttal flood this post with evidence of many kinds, on time scales from days to millennia or longer. I'll content myself with one pair of graphs spanning 50 years. They're from a 2007 report by Svensmark and the Institute's director, Eigil Friis-Christensen, and they're based on a European Space Agency project called ISAC. The carbon dioxide boys and girls would die for a match of cause and effect of this quality.

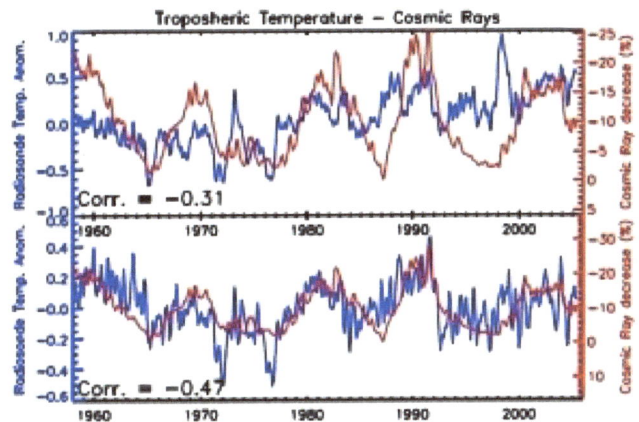

In this graph, cosmic ray intensity is in red and upside down, so that 1991 was a minimum, not a maximum. Fewer cosmic rays mean a warmer world, and the cosmic rays vary with the solar cycle. The blue curve shows the global mean temperature of the mid-troposphere as measured with balloons and collated by the UK Met Office (HadAT2).

In the upper panel the temperatures roughly follow the solar cycle. The match is much better when well-known effects of other natural disturbances (El Niño, North Atlantic Oscillation, big volcanoes) are removed, together with an upward trend of 0.14 deg. C per decade. The trend may be partly due to man-made *greenhouse* gases, but the magnitude of their contribution is debatable.

From 2000 to 2011 mid-tropospheric temperatures have remained quite level, like those of the surface, despite the continuing increase in the gases. This in flat contradiction to the warming predicted by the Intergovernmental Panel on Climate Change. Meanwhile the Sun is lazy, cosmic ray counts are high and the oceans are cooling.

Reference

Svensmark, H. and Friis-Christensen, E., "Reply to Lockwood and Fröhlich – The persistent role of the Sun in climate forcing," Danish National Space Center Scientific Report 3/2007.

Yet another trick of cosmic rays - In the climax to the Danes' experiments, cloud seeds flout the theories

Near to the end of the story that starts with stars exploding in the Galaxy and ends with extra clouds gathering, a small but important paragraph was missing till now. From experiments in Copenhagen reported in 2006 and reconfirmed in 2011 in Aarhus and Geneva (CERN, CLOUD), cosmic rays coming from old supernovas can indeed make molecular clusters a few millionths of a millimetre wide, floating in the air. But can these aerosols really grow nearly a million times in mass to be large enough to become "cloud condensation nuclei" on which water droplets can form – as required by Henrik Svensmark's cosmic theory of climate change?

Opponents pointed out that theoretical models said No, the growth of additional aerosols would be blocked by a resulting shortage of condensable gases like sulphuric acid in the atmosphere.

Not for the first time, an unexpected trick that Mother Nature had up her sleeve is revealed by experiment. The discovery is elegantly explained by a new way in which sulphuric acid forms in the atmosphere, as announced in a paper by Svensmark and two of his colleagues in Denmark's National Space Institute in Copenhagen, Martin Enghoff and Jens Olaf Pepke Pedersen. They have submitted it to Physical Review Letters. A preprint is available on **arxiv** here. http://arxiv.org/abs/1202.5156v1

Svensmark, Enghoff and Pepke Pedersen

A brief history. Last year's attempts to dismiss the Aarhus and CERN results as inconsequential for climate change didn't dismay the Danes. They knew there was something wrong with the current understanding because they had observational support for the whole chain from solar activity to cosmic rays to aerosols to clouds in the real atmosphere (Svensmark, Bondo and Svensmark 2009). In order to dig into the physics, they decided to rebuild, in the basement of the Space Institute, the eight cubic metre experimental chamber SKYII which six years ago was used as the CLOUD prototype chamber at CERN.

In the limelight of the atmospheric drama, sulphuric acid is one of the commonest of trace gases and very important for both the formation and the growth of aerosols. When the Sun rises in the morning, its ultraviolet rays convert sulphur dioxide, ozone and water vapour in the air into sulphuric acid molecules. These are continuously lost as they club together with further water and a little ammonia into very small molecular clusters. Nevertheless, the concentration of sulphuric acid rises to a peak and then diminishes as the Sun goes down in the evening.

A clue that something more is going on comes from the persistence all through the night of sulphuric acid at about 10 per cent of the daytime maximum. If these molecules too are continuously lost, they must be replenished by a chemical reaction that doesn't need ultraviolet light.

What did the new experiment called SKY2 show? Without going into technical details that you'll find in the paper, let's say that the primary result flatly contradicts the theoretical prediction that the infant aerosols couldn't grow up into cloud condensation nuclei.

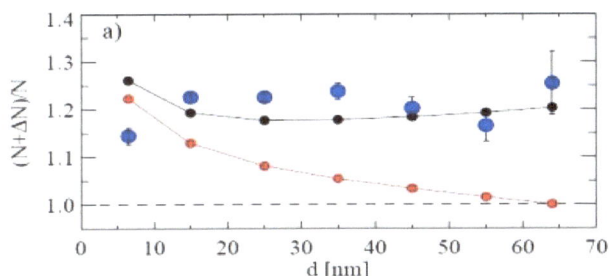

Here's a figure from the paper.

Molecular clusters grow over time, in the SKY2 experiment in Copenhagen. The horizontal axis is scaled in nanometres (millionths of a millimetre) and each blue point shows the relative number of clusters of that size before and after the experimental runs. Anything more than 1.0 means that growth has continued. In contrast, the red points illustrate a pessimistic prediction of previous theories, that growth should cease when the size passes 50 nanometres. On the other hand, the black curve running through the blue points shows what is to be expected if there is a continual supply of sulphuric acid. The persistent growth of clusters occurs only in the presence of gamma rays that simulate cosmic rays and set electrons free to influence the chemistry.

So what's the explanation? What new pathway supplies the sulphuric acid needed to keep the growth going? The Danes recall a suggestion in their 2006 SKY report that electrons can act like catalysts, being used over and over again to promote chemical action. In the new paper they say: A possible explanation could be that the charged clusters are producing additional [sulphuric acid] molecules from reactions involving negative ion chemistry of [ozone, sulphur dioxide and water], where a negative ion can be reused in a catalytic production of several [sulphuric acid molecules].

Depending on the concentrations of trace gases, several may mean dozens. And where do the electrons come from? They are liberated by cosmic rays raining down by night as well as by day. If the results of the experiment and these ideas are confirmed, there's an amazing payoff. The cosmic rays help to make the extra sulphuric acid that allows (1) a number of additional aerosols to form and (2) a larger number of aerosols to grow into cloud condensation nuclei. Without this second effect the aerosols would grow slowly and most of the extra aerosols would be lost before becoming large enough to seed clouds.

That ions liberated by cosmic rays promote a second pathway for forming sulphuric acid was already known from an experiment performed in Copenhagen in a collaboration with the University of Copenhagen and the Technical University of Tokyo (see the Enghoff et al. reference below). Depending on whether the sulphuric acid comes from ultraviolet action or is ion-induced, it has different signatures in the relative abundances of isotopes of sulphur. What's more, the number of molecules made by the ion route greatly surpassed the number of ions available, again implying reuse of the electrons in a catalytic fashion.

To summarize the latest paper, the Svensmark, Enghoff and Pepke Pedersen abstract reads:

In experiments where ultraviolet light produces aerosols from trace amounts of ozone, sulphur dioxide, and water vapour, the number of additional small particles produced by ionization by gamma sources all grow up to diameters larger than 50 nm, appropriate for cloud condensation nuclei. This result contradicts both ion-free control experiments and also theoretical models that predict a decline in the response of larger particles due to an insufficiency of condensable gases (which leads to slower growth) and to larger losses by coagulation between the particles. This unpredicted experimental finding points to a process not included in current theoretical models, possibly an ion-induced formation of sulphuric acid in small clusters.

Scandals of a political character engulf climate physics these days, but future historians may shake their heads more sadly over scientific negligence. Isn't it amazing that such a fundamental activity of sulphuric acid, going on over your head right now, has passed unnoticed since 1875 when cloud seeding was discovered, since 1996 when Svensmark found the link between cosmic rays and cloud cover, and since 2006 when the Danes suggested the catalytic role of electrons? Perhaps the experts were confused by the ever-present dislike of the role of the Sun in climate change.

So Svensmark and the small team in Copenhagen have had nearly all of the breakthroughs to themselves. And the chain of experimental and observational evidence is now much more secure:

Supernova remnants? cosmic rays? solar modulation of cosmic rays? variations in cluster and sulphuric acid production? variation in cloud condensation nuclei? variation in low cloud formation? variation in climate?

Svensmark won't comment publicly on the new paper until it's accepted for publication. But I can report that, in conversation, he sounds like a man who has reached the end of a very long trek in defiance of continual opposition and mockery.

References

Henrik Svensmark, Martin B. Enghoff and Jens Olaf Pepke Pedersen, "Response of Cloud Condensation Nuclei (> 50 nm) to changes in ion-nucleation", submitted for publication 2012. Preprint available at http://arxiv.org/abs/1202.5156v1

H. Svensmark, T. Bondo and J. Svensmark, "Cosmic ray decreases affect atmospheric aerosols and clouds", Geophysical Research Letters, 36, L15101, 2009

Henrik Svensmark, Jens Olaf Pepke Pedersen, Nigel Marsh, Martin Enghoff and Ulrik Uggerhøj, 'Experimental Evidence for the Role of Ions in Particle Nucleation under Atmospheric Conditions', Proceedings of the Royal Society A, Vol. 463, pp. 385–96, 2007 (online release 2006). This was the original SKY experiment in a basement in Copenhagen.

M. B. Enghoff, N. Bork, S. Hattori, C. Meusinger, M. Nakagawa, J. O. P. Pedersen, S. Danielache, Y. Ueno, M. S. Johnson, N. Yoshida, and H. Svensmark, "An isotope view on ionising radiation as a source of sulphuric acid", Atmos. Chem. Phys. Discuss., 12, 5039–5064, 2012

See http://www.atmos-chem-phys-discuss.net/12/5039/2012/acpd-12-5039-2012.html

Cosmic rays sank the Titanic

Full steam ahead for the real story of 20th Century warming

Although It seems a strange thing to celebrate, the Titanic Festival in Belfast, where the ship was built, will very soon mark the 100th anniversary of the liner's foundering on 15 April 1912 after hitting a south-wandering iceberg, with the loss of a multitude of passengers and crew.

Comparing the £100-million Titanic complex newly built in Belfast with the Guggenheim Museum in Bilbao, the travel writer Simon Calder has commented, "There is a great shipbuilding heritage. It is a divided city, but the Guggenheim is great on the outside but rubbish on the inside – unlike the Titanic building."

What's more, James Cameron's movie "Titanic" has been remastered in 3D for the centenary.

Time then for me to dig out some slides that I've used off and on in lectures since 1999 as an illustration of Henrik Svensmark's cosmic rays in action, controlling our climate. But first, just to show that I'm not being kooky, here's a graph from a 2000 paper by E. N. Lawrence of the UK Meteorological Office. "The Titanic disaster – a meteorologist's perspective" related iceberg abundance at low latitudes to a scarcity of sunspots. Steven Goddard recalls a much older

article, from the Chicago Tribune in 1923, that also linked icebergs with sunspots

http://stevengoddard.wordpress.com/2011/07/28/1923-article-linked-icebergs-with-sunspots/

The notion that the Sun is dimmer when there are few sunspots goes right back to William Herschel at the beginning of the 19th Century. The trouble is that the variations in solar brightness, as measured by satellites, are too small to explain the strong influence of the Sun on climate as recorded over thousands of years, and continuing into the 21st Century. That's where Svensmark's discovery of 16 years ago comes in, with the amplifier. Cosmic rays coming from the Galaxy are more intense when there are fewer sunspots and they increase the global cloud cover, so cooling the world.

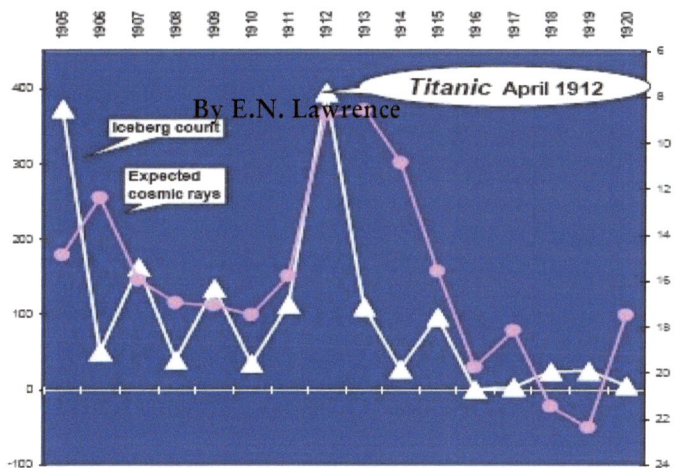

By E. N. Lawrence

Some preliminary comments before showing my own slides about cosmic rays and the fate of Titanic. Of course the disaster also involved several elements of shameful seamanship, but the fact remains that large icebergs abounded much further south than usual in the spring of 1912. Secondly, I prepared the slides so long ago that I can't recall the data sources. If challenged, I expect I could dig them out, and I do remember that the picture is from the Illustrated London News.

by Nigel Calder

It's a truly titanic idea, threatening disaster for the multitude who ignore the natural drivers of climate change, and shame for the misguided folk on the bridge who peer at computer screens instead of looking out of the window.

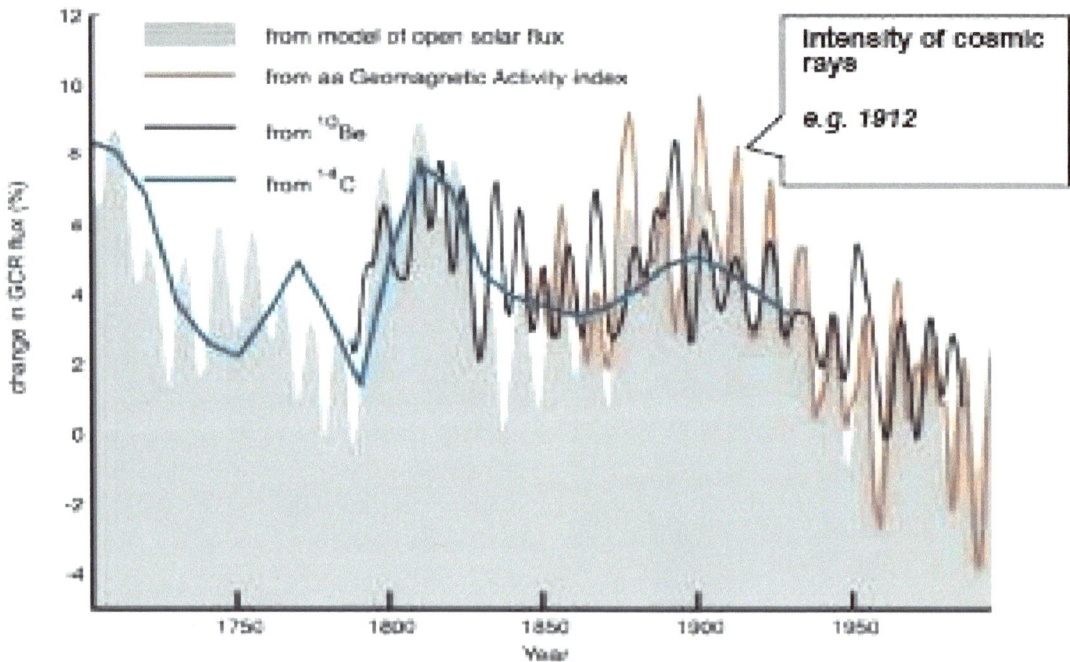

There was no direct recording of cosmic ray variations in those days. Indeed. Victor Hess was busy discovering them at that very time. So we have to make do with the geomagnetic activity index (called aa in the second slide, above) as an inverse indicator of cosmic ray influx, and with the counts of beryllium-10 and carbon-14, which are made by cosmic rays. Otherwise, the slides should speak for themselves.

The theme music of Cameron's film "Titanic" is entitled "Full Steam Ahead." Although the ship came to an abrupt halt, the same has not happened to Svensmark's theory. As plenty of other posts on this blog will show you, its bow wave keeps sweeping aside the attempts to falsify it. And fresh energy builds up more and more speed as all the pieces of the hypothesis fall into place, from quantum chemistry to the shape of the Milky Way Galaxy.

References

Simon Calder quoted:
http://www.belfasttelegraph.co.uk/business/business-news/titanic-site-to-exceed-all-expectations-says-expert-16114943.html#ixzz1nb8gmfMP

E.N. Lawrence, Weather (Roy. Met. Soc.), Vol. 55, March 2000.

See also this from NOAA
http://www.oar.noaa.gov/spotlite/archive/spot_sunclimate.html

Maybe It's Global Cooling We Should Be Preparing for

Tuesday, January 12, 2010

Repeat of a post from 2005 with some new comments

This is a repeat of a paper I used in a lecture given to the South Bend Chapter of the Engineering Society in 2004. I also published in 2005 in a blog titled, "Maybe It's Global Cooling We Should Be Preparing for." The article speaks for itself in pointing out the errors in the very first promotion of "global warming" including Al Gore's book. It should be quite obvious to even the dimmest lightbulb out there that this is a monstrous scam perpetrated solely for money and power on an international scale. Sadly it has blackened the reputation of good science in the eyes of so much of the public who are not as ignorant or stupid as those greedy politicians think. It is my firm belief that as the truth about the realities of climate change finally comes out, the damage to the respect the scientific community worked so hard to earn will be monstrous. Those scientists who have joined politicians and abandoned their scientific principles to jump on the global warming bandwagon without adequate investigation will surely suffer in the eyes and opinions of the public. This fiasco will take decades to correct if indeed it is correctable.

Those proclaiming the dangers of global warming would do well to study what real climate scientists have discovered about the last few thousand years.

The paper begins:

Many climate scientists believe it is quite possible that the changing gulf stream indicates the likelihood of a return of conditions that brought on the *Little Ice Age* from mid 14th century 'til the start of the twentieth century. Should the gulf stream stop flowing, as many scientists believe it did during the *Little Ice Age*, the increased carbon dioxide in the atmosphere may actually temper the temperature drop and make another such climatic event less damaging although by a minuscule amount.

Most of the negative information about global warming comes from computer modeling which, in the past, has been consistently wrong. Temperature records of the last several thousand years indicate repeated run ups and declines in average global temperatures far greater than we are currently experiencing. The following article indicates global temperatures for numerous centuries from 800AD to 1300AD averaged far greater than those of the twentieth century.

The real problem is that politicians and some scientists, who are promoting self-serving agendas, concentrate solely on data supporting their chosen position and ignore data that disproves it. Most scientists preaching alarm about global warming are not in the field of long range world climate. As a matter of fact, most scientists working in that field believe many of the observed temperature fluctuations fall well within historical limits of fluctuations.

The truth is that attempts at computer modeling of worldwide climate changes and weather in either short or long range are far too inaccurate for dependable results. To have meaningful results they would have to be able to make accurate predictions about a future hurricane, before it ever started. Current modeling can hardly predict even a few day's future movement and power of an existing hurricane with any degree of accuracy. The technical limits and variables facing such modeling is far far beyond our present technical capabilities.

20th century not warmest, researchers find

Coral reefs are sensitive to a variety of environmental changes. Smithsonian astronomers Willie Soon and Sallie Baliunas reviewed more than 200 studies of coral, glacier ice cores, tree rings and other indicators to trace changes in the world's climate over the past millennium. They found that the 20th century is NOT the warmest of the past thousand years. (Credit: David A. Aguilar, Harvard-Smithsonian Center for Astrophysics)

11:30 a.m., April 15, 2003--A review of more than 200 climate studies led by researchers at the Harvard-Smithsonian Center for Astrophysics has determined that the 20th century was neither the warmest century nor the century with the most extreme weather of the past thousand years.

The review, which included work by David R. Legates, director of the University of Delaware's Center for Climatic Research, also confirmed that the *Medieval Warm Period* of 800 to 1300 A.D. and the *Little Ice Age* of 1300 to 1900 A.D. were worldwide phenomena not limited to the European and North American continents.

Legates said the paper argues against a recently espoused view formulated by Michael Mann of the University of Virginia and his colleagues that global air temperatures remained fairly constant from 1000-1900 A.D., then increased dramatically in the 20th century.

"Although [Mann's work] is now widely used as proof of anthropogenic global warming, we've become concerned that such an analysis is in direct contradiction to most of the research and written histories available," Legates said. "Our paper shows this contradiction and argues that the results of Mann . . . are out of step with the preponderance of the evidence."

According to the paper, while 20th-century temperatures are much higher than in *the Little Ice Age* period, many parts of the world show the medieval warmth to be greater than that of the 20th century.

Smithsonian astronomers Willie Soon and Sallie Baliunas, with co-authors Legates and Craig Idso and Sherwood Idso of the Center for the Study of Carbon Dioxide and Global Change, compiled and examined results from more than 240 research papers published by thousands of researchers over the past four decades. Their report, covering a multitude of geophysical and biological climate indicators, provides a detailed look at climate changes that occurred in different regions around the world over the last thousand years.

"Many true research advances in reconstructing ancient climates have occurred over the past two decades," Soon said, "so we felt it was time to pull together a large sample of recent studies from the last five to 10 years and look for patterns of variability and change. In fact, clear patterns did emerge showing that regions worldwide experienced the highs of the *Medieval Warm Period* and lows of the *Little Ice Age*, and that 20th-century temperatures are generally cooler than during the medieval warmth."

Soon and his colleagues concluded that the 20th century is neither the warmest century over the last thousand years, nor is it the most extreme. Their findings about the pattern of historical climate variations will help make computer climate models simulate both natural and man-made changes more accurately, and lead to better climate forecasts especially on local and regional levels. This is especially true in simulations on time scales ranging from several decades to a century.

Studies of stalagmites and tree rings can yield yearly records of temperature and precipitation trends.

Researchers drill small cores to obtain samples. NOTE: Tree ring growth is not a valid measure of temperature for the last century since the increase of atmospheric CO_2 enabled trees to grow faster even more than increases in temperature.

Historical cold, warm periods verified

Studying climate change is challenging for a number of reasons, not the least of which is the bewildering variety of climate indicators–all sensitive to different climatic variables, and each operating on slightly overlapping yet distinct scales of space and time. For example, tree ring studies can yield yearly records of temperature and precipitation trends, while glacier ice cores record those variables over longer time scales of several decades to a century.

NOTE: Tree ring data is extremely unreliable as an indicator of air temperatures since they vary in width with changes in rainfall and concentration of CO_2. Correcting for these factors is virtually impossible

Soon, Baliunas and colleagues analyzed numerous climate indicators including: borehole data; cultural data; glacier advances or retreats; geomorphology; isotopic analysis from lake sediments or ice cores, tree or peat celluloses (carbohydrates), corals, stalagmite or biological fossils; net ice accumulation rate, including dust or chemical counts; lake fossils and sediments; river sediments; melt layers in ice cores; phenological (recurring natural phenomena in relation to climate) and paleontological fossils; pollen; seafloor sediments; luminescent analysis; tree ring growth, including either ring width or maximum late-wood density; and shifting tree line positions plus tree stumps in lakes, marshes and streams.

"Like forensic detectives, we assembled these series of clues in order to answer a specific question about local and regional climate change: Is there evidence for notable climatic anomalies during particular time periods over the past thousand years?"

Soon said. "The cumulative evidence showed that such anomalies did exist."

The worldwide range of climate records confirmed two significant climate periods in the last thousand years, the *Little Ice Age* and the *Medieval Warm Period*. The climatic notion of a *Little Ice Age* interval from 1300 to 1900 A.D. and a *Medieval Warm Period* from 800 to 1300 A.D. appears to be rather well-confirmed and widespread, despite some differences from one region to another as measured by other climatic variables like precipitation, drought cycles, or glacier advances and retreats.

"For a long time, researchers have possessed anecdotal evidence supporting the existence of these climate extremes," Baliunas said. "For example, the Vikings established colonies in Greenland at the beginning of the second millennium that died out several hundred years later when the climate turned colder. And in England, vineyards had flourished during the medieval warmth only to shrink or perish during the *Little Ice Age*. Now, we have an accumulation of objective data to back up these cultural indicators."

Glacier ice cores record temperature and precipitation trends over longer time scales of several decades to a century. (Credit: Lonnie Thompson, Byrd Polar Research Center, The Ohio State University)

The different indicators provided clear evidence for a warm period in the Middle Ages. Tree ring summer temperatures showed a warm interval from 950 A.D. to 1100 A.D. in the northern high latitude zones, which corresponds to the " *Medieval Warm Period*." Another database of tree growth from 14 different locations over 30-70 degrees north latitude showed a similar early warm period. Many parts of the world show the medieval warmth to be greater than that of the 20th century.

The study–funded by NASA, the Air Force Office of Scientific Research, the National Oceanic and

Atmospheric Administration, and the American Petroleum Institute, will be published in the Energy and Environment journal. A shorter paper by Soon and Baliunas appeared in the Jan. 31 issue of the Climate Research journal.

End of paper - beginning of part of the blog added to the end of the paper in 2005.

It is quite obvious that the growing "global warming" movement is not based on good science. Good science does not ignore data that disproves an hypothesis and look only at that which does. Good science does not ignore data points on graphs used to promote a concept just because those points do not agree with a particular agenda. Good science does not refuse to look at or consider basic math, chemistry, and physics which do not bear out the hypothesis. Good science does consider all the data, positive, negative, and anywhere in between. Good science does not place the opinions or theories of any person or group above the provable facts no matter what the reputation of the person or group. Good science does not completely ignore the fact that earth has now been in a cooling trend for the last sixteen years. Good scientists are very skeptical, especially so where politics or politicians are involved. It has been said and wisely so that one should not ask a scientist to disprove a paranormal claim, ask in stead a magician.

I must say, I am not an expert on climate, but I have looked at and studied data, graphs and charts about climate, some going back more than a hundred thousand years. This information is from many sources and has the benefit of the test of years. To this discerning eye, those charts and graphs point to a major cooling event, possibly even the beginnings of another "Ice Age" in the very near future. These type events take place over millennia, even centuries, and can be the cause of huge changes in the ecology of much of the earth. The last ice age ended around ten thousand years ago after nearly sixty thousand years of holding most of the current temperate zones in an icy, arctic grip. North America was buried with up to a mile of ice as far south as the Ohio river and the Rockies were buried all the way to Mexico. Life in this frigid land was tough for mammals and impossible for all cold blooded creatures. Trees and flowering plants were non existent anywhere north of the snow line. Europe and Asia were similarly affected with all mountains buried as well. Glaciers even flowed into the Mediterranean at times. Sea levels were much lower with land bridges connecting many places now under water. It was a very different world.

Many dramatic climate changes have occurred over very short time periods, decades or even less. In fact many scientists now believe sudden changes of a few decades or even a few years are the rule rather than the exception.

See this site for more:

http://www.esd.ornl.gov/projects/qen/transit.html

After examining the data at hand it is my conclusion that we are near or at the end of the current interglacial period. For this reason and in spite of all the noise about the effects of CO_2, I am firmly convinced we are on the verge of a major change to a much cooler earth. Whether it is another *Little Ice Age* that will last a few centuries before it warms up to a period like the present (a common happening) or it's another long and devastating Ice Age like the last one that buried much of the northern hemisphere under as much as a mile of ice, remains to be seen. If it's the former most of us will feel its effects. If it's the latter, well, you better consider moving to a much warmer climate—now! The most likely culprit? Earth's constantly changing position and distance from that small, variable star we call the sun. That's been messing with our climate for billions of years and there's not much we can do about it but worry and adapt. Sorry about that all you global warming profiteers. Your gold mine will, I fear, soon pan out to nothing.

End of 2005 blog

The erroneous claims about CO_2 and its effect on air temperature are very troubling. Even doubling the amount of CO_2 in the atmosphere would have a negligible effect on these huge climate shifts. Do the math, using the gas laws, and the relative net energy retention coefficients of the various atmospheric gasses, figure out how much of a temperature rise this change in the amount of CO_2 would bring about. It's simple, basic first year college chemistry, maybe even high school. The results show it's akin to dumping a bucket of water into even a moderate sized body of water like Lake Erie, or even the local fishin' hole. That's a far cry from the claims of drowned lands, terrible storms, violent weather, and worldwide calamity we hear constantly from the media and politicians eager to grab some of your money. I have a hard time believing the public are so gullible.

Let me reiterate a prediction I made first in 2004 and posted to my blog in 2005.

We are probably on the verge of changes to a cooler Earth in the next few centuries. The *Little Ice Age* descended on Europe (1400-1900 AD) after the *Medieval Warm Period* (800-1300 AD) which was considerably warmer than the current warming period (1900-the present, 2004). Data shows that solar activity is beginning a typical change to less energy from the sun reaching the earth. This set of conditions combined with several other cyclic changes, usually heralds a cooler Earth like the *Little Ice Age* or even an extended major cold and ice event like the Younger Dryas (12,800 BP to 11,600 BP - Before Present) when the earth returned to full "Ice Age" conditions for more than a millennium. Both the start and end of the Younger Dryas event were very abrupt—five to fifty years. There are several competing theories as to the direct causes of this event which coincided with major megafauna extinctions. There is also evidence that shows the temperature drop that ushered in the *Little Ice Age* during the 14th century may have occurred in a single season. That could explain why the Norse settlement on Greenland, active and successful for at least 500 years was abruptly wiped out when the temperature dropped suddenly.

Here's an excerpt from the Saturday, March 28, 2009 posting in this blog:

The world's climate system is infinitely more complex than a single hurricane season. It moves in cycles and eddies that run from seconds to millennia. About forty years ago some climate pundits feared we were heading into global cooling and needed to prepare for a drier, cooler time with lower sea levels. According to many scientific studies of past frigid periods we are past due for the onset of the next ice age. Hubert Lamb of the UK Met Office dominated the 1961 UN meeting on global cooling. A founder of the Climate Research Unit at East Anglia, he was one of the world's top climate scientists. He warned that people had become complacent about climate at a time when population growth, cold, and drought could seriously damage their food supplies. (The Norse in Greenland perished of starvation after five hundred successful years when the *Little Ice Age* destroyed their crops.) In historic times the climate has veered from warmer than the present, the *Medieval Warm Period*, to the much colder conditions of the *Little Ice Age* from which we may still be emerging. Evidence shows that much of the Sahara and the Middle East held lush vegetation and crop land twelve thousand or so years ago while northern Europe and America were covered with up to a mile of ice during the Younger Dryas event.

On the next page, the IPCC's crucial *hockey stick* graph, shows a dramatic rise in temperatures over the last 40 years of the 20th century - but what happens to the green line of direct temperature measurements? The two lower details show how the data was misrepresented in order to prove what the powers that be in the IPCC want to prove. The facts bear no relationship to their misleading graphs.

This is but one small part of the scam exposed by those hackers who broke into the East Anglia Climate Department computers.

Doesn't that kind of salacious activity throw a great deal of doubt on the reliability and efficacy of all manner of "scientific" data and pronouncements? Doesn't that reflect on the reputation of the entire scientific community? Can anyone still have confidence in those bits of "scientific" information released to the public, especially through our "unbiased" media? Can we believe anyone anymore or are all these people like our politicians? That is indeed a sorry state of affairs driven by greed and political ideology, not science.

"Climategate" is the name ascribed to the scandal revealed by the leaked emails from the University of East Anglia's Climatic Research Unit. The reason this was so shocking is that the authors revealed were a small group of scientists who for years were influential in pushing worldwide alarm over supposed global warming because of increased atmospheric CO_2. They were very influential in convincing the UN's Intergovernmental Panel on Climate Change (IPCC) to warn the world of exaggerated dangers of Anthropomorphic Global Warning and propose huge taxes on CO_2 emitters, primarily the US. Remember the Kyoto protocols?

There were two key sets of data the IPCC used in its reports. The IPCC's main scientific contributors claimed these global temperature records are the most important of all of the sets of temperature data the IPCC and many governments rely on to predict future global temperatures. They predict the world will reach catastrophic temperature levels if trillions of dollars are not provided governments to use to avert it.

A group of American and British scientists were organized to promote the picture of atmospheric temperatures predicted by Michael Mann's "hockey stick" graph which was manipulated to indicate that global temperatures have risen sharply after 1,000 years of decline, and that they have risen up to the highest levels in recorded history. They completely ignored the long-accepted Mediaeval Warm Period when temperatures were higher than they are today. During this 500 year period the Norse lived and raised cattle on farms, built cities, churches and schools in Greenland only to starve and die out completely when the weather turned cold for the little ice age. Their carefully crafted and misleading graph became their proof of the man-made global warming movement, a political movement which has almost become a religion for its adherents.

This is the reality of the scam they pulled as revealed by the famous hackers.

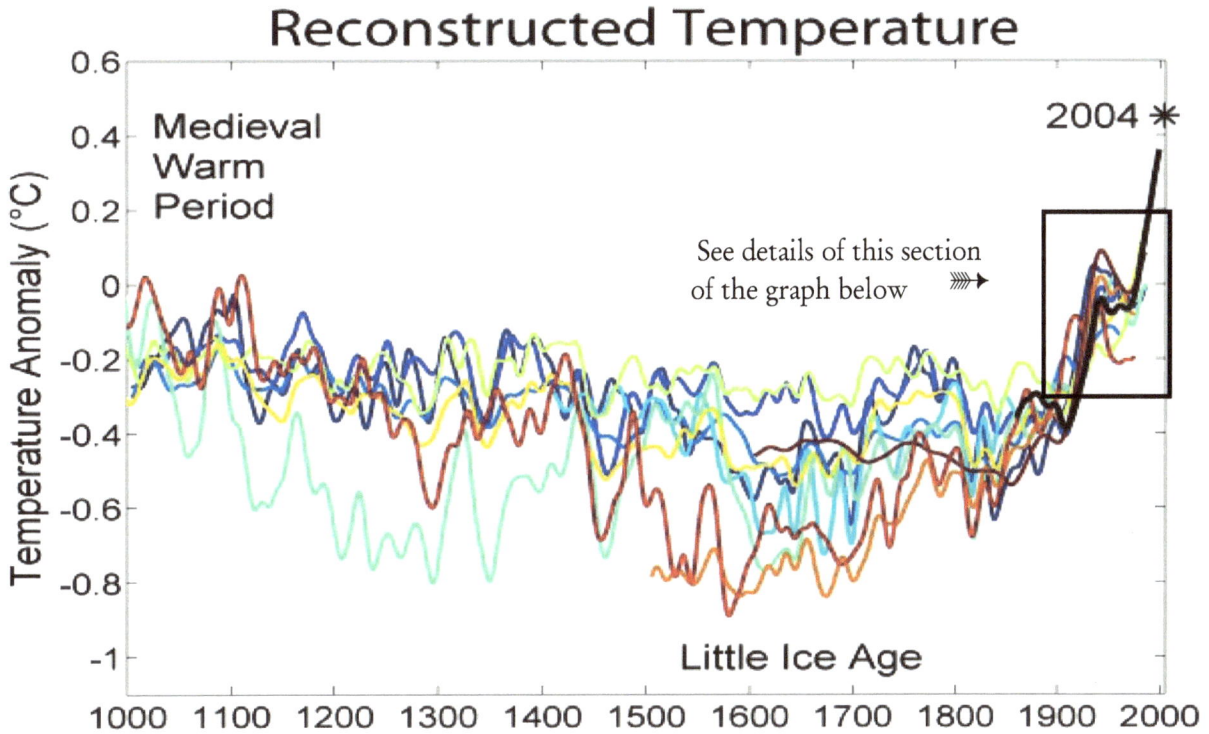

Reconstructed Temperature

Medieval Warm Period

2004 ✳

See details of this section of the graph below ⋙▶

Little Ice Age

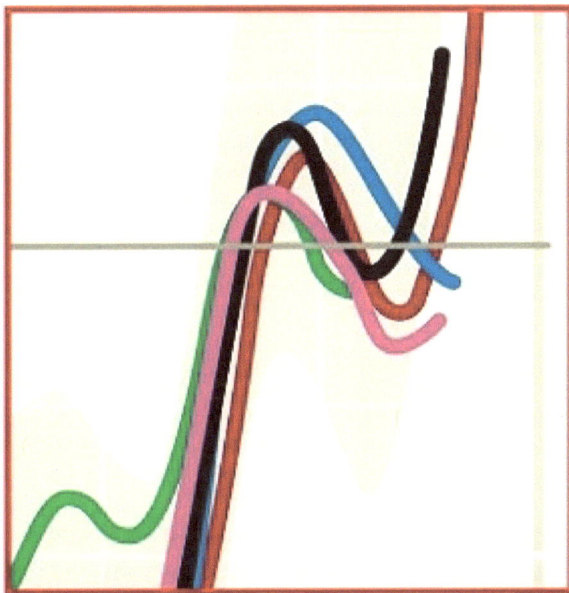

Blowing up the graph shows it disappears in 1961, artfully hidden behind the other colours

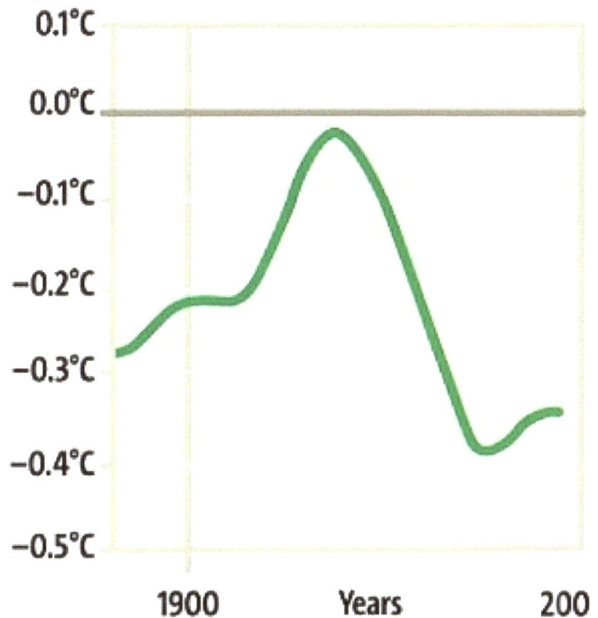

The reason? Because this is what it shows after 1961: a dramatic decline in global temperatures …

Bully for Global Warming and a Warmer Climate!

The other day I got to thinking, don't the anthropogenic global warming fanatics have one important fact upside down? Despite all the hullabaloo, the promoters, and the deniers, *Anthropogenic Global Warming*, (AGW) could be the best thing to happen to our planet in many millennia and definitely more good than bad.

One scholarly tome about ice ages (like the most recent one we may still be coming out of) tells us in no uncertain terms that global warming would be quite advantageous for life on earth, even humans. The benefits would far outweigh the losses. It would certainly be a lot better than another ice age. So I say, quit wringing your hands and crying of doom. Cancel those carbon taxes. Global warming could be one of the best things to happen since the last ice age ended.

While there will certainly be many winners, there are bound to be a few losers. Sorry, New York and LA, but you'll have to move a bit inland. I'm terribly sorry about drowning Florida, but no change is ever all good. Besides, the lush semi tropical shores of Labrador will be habitat for former Florida flora and fauna while the lowlands of Georgia, the Carolinas, and the rest of the southern states all the way to the West Coast will be lush tropical paradises. Manatees on the Labrador coast and alligators in Canada would be a big change for the better. That's certainly a much prettier picture than that painted by the doom sayers of the church of **catastrophic** global warming. Al, eat your heart out!

The one glaring error the believers in **catastrophic** global warming make is saying that it is catastrophic. The fact is it would be almost entirely **beneficial** to mankind and to all life on the planet, and substantially so. Warm periods, such as the *Medieval Climate Optimum*, were historically beneficial for civilization. In contrast, corresponding cooling events, such as the *Little Ice Age,* were uniformly bad news. Up to a point, the warmer our planet gets, the more life it supports and the more variety that life has. In contrast, the colder it gets, the less life and the fewer species exist. Read that as less food for humans. All fossil records support that reality as far back as the Permian period, 500 million + years ago.

We have very good evidence that a warmer planet has far more benefits than negatives. It is well known from several sciences including botany, zoology, and paleontology, that almost all life has fared much better during warm periods on the planet than during colder periods. The amount and diversity of life were severely diminished during the ice ages, expanded as the glaciers receded, and were much greater during periods warmer than the present. The borders between tropical weather and temperate, as well as between temperate weather and arctic, move toward the polar regions as the planet warms. The quantity and variety of life increases when temperate areas gave way to tropical, and arctic gave way to temperate.

The East Anglia revelations damage the credibility of scientists: The pointed dishonesty of the weather *scientists* as revealed by the information *hacked* from university records is appalling. The truly sad and destructive inferences not missed by the public are the sinister shadows of doubt now thrown over the reports of the entire scientific community. The very integrity of those engaged in scientific research and reporting has been called into question everywhere.

While honest mistakes and misjudgements have frequently caused doubts about *scientific* reports and publications, dishonesty has been an extremely minor issue. Now abject dishonesty including manipulation of computer code used in weather simulations on super computers has become evident in what is probably one of the most powerful *science* controversies of current times. This is without precedent and is deeply damaging to the entire scientific community. As a result, scientists will face an increasingly skeptical and doubting public for a very long time. The people have been lied to by a few scientists they trusted, and they certainly do not like it. Because of this, all of science has now come under suspicion, and that is a real shame.

The one glaring error the believers in **detrimental** global warming make is that it is **catastroph**ic. The fact is it would be almost entirely **beneficial** to mankind and to all life on the planet, and substantially so. Warm periods, such as the *Medieval Climate Optimum*, were historically beneficial for civilization. In contrast, corresponding cooling events, such as the *Little Ice Age*, were uniformly bad news. Up to a point, the warmer our planet gets, the more life it supports and the more variety that life has. In contrast, the colder it gets, the less life and the fewer species exist. All fossil records support that reality as far back as the Cambrian period, 500 million + years ago.

We have even more evidence that a warmer planet has far more benefits than negatives. For obvious reasons, *global warming* advocates never consider that very significant fact. It is well known from several sciences including botany, zoology, and paleontology that most all life has fared much better during warm periods on the planet than during colder periods. The amount and diversity of life were severely diminished during the ice ages, expanded as the glaciers receded, and were much greater during periods warmer than the present. During warmer periods, the borders between tropical weather and temperate, as well as temperate and arctic on our planet, move toward the polar regions. As a result, the quantity and variety of life increases when temperate areas gave way to tropical, and arctic gave way to temperate. This is a well-established and easily understood fact.

Compare the number and variety of life as it now exists in the tropics, in the temperate zone, and in the arctic. Imagine what would happen if the tropics expanded and the arctic zone shrunk. Would that be beneficial to life or detrimental? Obviously it would be beneficial. I'll wager few people ever even considered that fact. There is, in fact, a greater number and variety of species in a square mile of Amazon jungle than in the entire Arctic and Antarctic regions of our planet. For obvious reasons, the high priests of the harmful global warming movement would never mention this fact.

A case in point in human experience: The Vikings established a thriving community of farms, villages and churches in Greenland that lasted for at least 500 years during the *Medieval Warm Period* that was somewhat warmer than today. There is even evidence they established colonies in North America (Vinland?) during the same period. Then, when the *Little Ice Age* happened and arctic weather moved south over Greenland, they starved and disappeared leaving their fields, farm buildings, homes and churches where they stood. Greenland's tiny Inuit population persisted until the present. In the eighteenth century, the Danes reestablished a small colony where one of the Viking settlements had been. That colony has survived only with imports of substantial amounts of food, building materials, and hay for animals. For all practical purposes, the Vikings are still the only self-sustaining population of Greenland, other than a few Inuit hunters. Present-day greenlanders now place their hopes for the future on utilization of vast, untapped mineral resources

Suppose the worst claimed by the gurus of *global warming* actually happened and sea levels rose a hundred feet. During this period, the tropics would expand 500 to 1,000 miles, and the temperate zones would move toward the poles about the same distance, Should this happen, large areas of northern Canada and Siberia, among others, would become much more hospitable to life of many kinds—temperate forests and fertile farmlands. Since it is well known that warmer periods brought wetter weather patterns, many deserts would become lush and green. Talk about a green revolution—that would be a real one! Southern California, Arizona, New Mexico and West Texas valleys could become tropical jungles. The same could happen to the deserts: the Sahara, the Gobi, and even the Australian. Wouldn't that be a kick? Who knows, with the ice gone, Greenland might even become green and habitable. One thing for sure, the Arctic ocean would become an open shipping lane. The overall benefit to life on the planet would be huge. So we would lose some land and a few cities along the coast and many islands would shrink or disappear, big deal! Since it would happen gradually, the people and much of their property could easily be moved elsewhere.

Oh yes, the polar bears and penguins. Those cute little birds and those big, soft and cuddly bears so painted by the brush of the media to appeal to children. Actually, they are both predators (as are we humans, even children). Penguins kill and eat millions of poor little fishes. And polar bears? They kill and eat many seal pups every year. Think of those poor, cute little seal pups being brutally dragged from their lairs, slaughtered, and eaten. Also, they would gladly kill and eat you if given the opportunity and they were hungry. Of course, they much prefer tasty seal pups. Should the planet warm substantially, the environment of their habitat would change and they would probably go extinct. Sad, is it not?

The expanded tropics would become habitats for many individuals of hundreds of new species, many just as loveable as penguins and polar bears. Think more cute little monkeys and colorful new birds in lush jungles teeming with life. Think also about the manatees, armadillos, big cats, alligators, crocodiles, snakes, lizards, spiders, and thousands of other creatures you never see in the arctic. They would be found in countless numbers in temperate areas that replace arctic tundra and permafrost, as well as in the tropics that replace temperate areas. Compare those habitats with the cold, bleak, windswept arctic of snow and ice, supporting but a few creatures in hundreds of square miles.

And how about this for you or me, a sunny, warm beach with palm trees and colorful birds compared with a frigid, windswept snowy plain, or maybe even farmland on ground once covered with a mile of ice in Greenland. That same ice once covered almost all of Canada and most of the northern United States and Europe. Global warming removed that ice and turned that land into woods, farms, vineyards, and varied wild habitat? Considering those known and obvious facts, does global warming seem so terrible?

Remember those foreboding warnings about nuclear winter, with the cold killing everything? How about the same thing caused by volcanic eruptions or an enormous meteor strike? Contrast that with a more temperate climate where the arctic ice melts and everything gets warmer. Palm trees covering Labrador, temperate forests all over Greenland, and tropical rivers crossing the Sahara. That is what global warming would probably produce long-range. It has in the past, long before man existed, and could certainly do so again in the future, probably long after man has gone extinct.

A Research Scientist at East Anglia Speaks out

Another shocker the liberal controlled media will not report relates to new information on the benefits of increased atmospheric carbon dioxide. The following quote comes from a scientist and professor from a well-known university whose specialty is studying the effects on plant life of increased carbon dioxide in the atmosphere. It was a comment made to a blog I posted of this *Bully for Global Warming* chapter. He does not want his identity revealed for obvious reasons. (Peer pressure and access to government grant money.) He wrote about the benefits of increased CO_2 in a comment to my BLOG.

"What your very good essay doesn't much focus on is the tremendous increase in plant growth that has already been afforded by the increase in CO_2. This is generally calculated to be in the range of 15% to 40%, depending on the species (and whether it's a C4 or C3 type of photosynthesis) -- which means at minimum about a 1/7th increase in plant productivity.

"This means in turn that something like a billion humans are being fed by the EXTRA crop growth from increased CO_2. So the contributions of Man, through fossil fuel, land use changes and agriculture are having a very positive effect.

"Think of the harm -- the starvation -- if the extra CO_2 could be instantly made to go away, as the catastrophists apparently wish. It is not a pretty scenario at all. By the way, the programming mischief revealed in the CRU release is, if anything, worse than the emails themselves."

http://community.livejournal.com/global_warming/11557.html

Level Head - November 26, 2009 12:45 P.M.

In correspondence with him since then, we agreed that there was a dearth of research, even curiosity, about beneficial effects of increased CO_2 in the atmosphere. We each wondered why this was so. I suggested that was because positive information about increased CO_2 would have a negative effect on the AGW gravy train. Politicians and the media would not allow such information to be made public knowledge. They might lose one of their most

powerful political tools, one that provides them with scads of money.

After realizing these facts, maybe Al Gore will write a new book titled, *A Beautiful Truth* that will tell the facts about global warming. Oh, but that wouldn't work with his liberal agenda, would it? The new American liberal politics trumps facts, science, and all other forms of truth. Shades of the old Soviet Union. This is especially true of all information showing the benefits of increased carbon dioxide and of global warming, whether related or not. Certainly, the degree of their relationship has not been established by anything other than anecdotal evidence.

So I say, burn that oil, burn that coal, and pump out that carbon dioxide! Let's heat up the planet and head for those tropical beaches in Labrador, visit oak forests and farms on Alaska's north slope, or rent a seaside villa on the warm shores of Greenland? Sounds fantastic to me. It certainly doesn't sound as bad as the AGW profits of doom predict.

❖   ❖   ❖

An Exchange With a "Real" Scientist (so she says)

Tuesday, December 8, 2009

This is an exchange between a lettered scientist from academia and the author, a private sector engineer, about atmospheric CO_2, climate change, and whether they are related, or are harmful or beneficial. Specifically, it is in response to the "Bully for Global Warming" November 25, 2009 section posted earlier in this blog. The names are changed to protect the guilty. Incidentally, her degree is quite remote from climate science and bears little in common with the complex systems that comprise climate. Methinks perhaps that in the field of climate science, her political ideology trumps her scientific reasoning and conclusions. Shades of the sad histories of the soviet scientists whose conclusions ran counter to the Communist party line.

The scientist writes:

"My dear dear Howard, no one is denying that temperatures were much warmer in past eons. That is not the point. The point is the rate of change in climate, which cannot reasonably be attributed to anything but an enhanced *greenhouse* effect due to the addition of *greenhouse* gases to the atmosphere (which even your favorite skeptics can't deny) from fossil-fuel burning. Your friend the (real) scientist, Crista"

NOTE: That statement about the rate of change is patently untrue. It is refuted by most of the content of this book, *Climate & Much Worse Dangers We Ignore*. My only favorite skeptic is myself and I do deny the accuracy of that statement. Besides, all true scientists are (or should be) born skeptics of any theory lacking in hard science proofs. Unlike some individuals, I do not let the dictates of others determine my opinions or understanding. Provide me the evidence. I will make my own decisions as to what the evidence does or does not prove. Don't tell me, "Groups of scientists believe it, so that makes it true." Incidentally, describing herself as a (real) scientist implies that scientists that do not support AGW are not (real) scientists.

I responded

My dear, dear Crista: You miss the point of my blog completely. All I am saying is that any normal increase in global temperatures from any cause whatsoever is much more beneficial to life including man than detrimental. So why should we make such a huge effort to stop something that is beneficial?

All that effort is accomplishing is to divert vast amounts of money into the hands of politicians who, by the way, dole out small parts of their spoils to academia. These politicians could care less about real dangers, even menaces, to the planet which are

considerable. I even wonder if the emotional attachment of so many to the AGW phenomena is because they haven't the guts or intellect to face the real problems which are far more dangerous. Or do they just like the money?

A threat and a menace are not the same. A threat may be weak or empty and thus can be ignored. A menace on the other hand must be dealt with or there will be a catastrophe. Menaces come in many sizes. AGW, even if it is real, is an extremely tiny menace by any measure when compared with the real menaces we face. The global warming movement with its growing and soon to be massive transfer of wealth from the private sector to governments, is a real menace to the well being of most ordinary citizens of the free world, particularly America.

I disagree entirely with the statement that **the present** *rate of change in climate, which cannot reasonably be attributed to anything but an enhanced greenhouse effect due to the addition of greenhouse gases to the atmosphere from fossil-fuel burning.* That rate is still quite slow compared to past dramatic climate changes.

My only "favorite skeptic" is my own mind. I pay only passing attention to others of any kind. Unlike those in the closed world of academia, I am not beholden to the prejudices and agendas of grant committees, peers, or political/financial benefactors. I am free to search where I want, say what I want, and write what I see and understand without fear of damage to my prestige, my career, or my financial condition. No one has a hold on me of any kind. Any supposed factual statements, calculations, or proofs I offer are subject only to correction by better or more accurate statements, calculations, or proofs. My opinions? Ideally, they carry as much weight as anyone's. Realistically, they only carry as much weight as that of those who agree with them.

No, I am not what lettered academicians would call a (real) scientist, but I consider myself one who believes in and abides by the scientific method, and I am a (real) PE (professional engineer) who spent a great deal of time in training, study, and application of the gas laws ($PV = nRT$ etc.), the laws of physical chemistry, and the laws of thermodynamics. As a chemical engineer, I studied all of these extensively. From these studies I learned to understand the way any gas heats up in sunlight and cools in the blackness of space by black body radiation. It is a factor of the absorption and emission of heat energy by the gas molecule and is not very difficult to calculate. These laws lie at the very heart of the theory of AGW. Their factual accuracy cannot be denied. Their application would prove conclusively the range of temperature change increasing amounts of CO_2 in the atmosphere does and would bring about. Why is it that this mathematical and physical proof is never used, conducted, or mentioned? The answer is quite simple. It would counter all of the *consensus understanding of human-induced, global climate change, which is a robust hypothesis based on well-established observations and inferences.* (In plain English, it is merely *the opinion* of one group of humans.)

If you, Crista, or anyone else can provide me those calculations, and they are accurate, not skewed, and they prove your position is correct, I will immediately change my opinion and my writing.

In an older part of this same blog titled "Earth Hour" and under the heading "Some practical information about the processes acting on our atmosphere" there is some information about these gas laws and some calculations that are written in non technical terms. If they are in error, prove it. I have written to many organizations and even a few institutions of higher learning asking for the math and physics that proves AGW has anything but a negligible effect on atmospheric temperatures compared with the many other factors that have well-studied effects. I have applied the physics and done the math and was asking for confirmation or contradiction. While I received several references to

published articles (all of which had statistical and anecdotal information, but no math or physics) and several comments suggesting I "leave such questions to the professionals," I received not a single response with any answers to my physics or math questions or to my comment about the grossly erroneous *greenhouse* analogy that has become so firmly ensconced in the minds of the public. I did respond rather pointedly to those who wrote, "leave such questions to the professionals," saying things about credentials being suspect and not seeing the forest for the trees. I also said that if they didn't have the answers to my physics and math questions, they should admit it. Fat chance!

Crista, perhaps one of your colleagues could apply some real physics and math and refute or confirm the paragraph I mentioned.

"Show me the beef!" I don't care about the sizzle.

Howard

She responded:

"Dear Howard, Yes, there is some potential good to come from climate change, but on balance it will not be good, based upon everything that I have read and studied. I don't really want to debate this with you; the attached is a report from the US Climate Change Science Program that I worked on, released this summer. I think it is useful information that is based upon many years' worth of solid research. Take it or leave it; it is what it is. Happy birthday, by the way. Crista."

NOTE: The publication is *Global Climate Change Impacts in the United States*, "A State of Knowledge Report from the US Global Change Research Program" according to the group that produced it. It is an impressive 188 page (8.5" x 11") document with a lot of interesting and some disturbing information about what we should do to prepare for higher temperatures. It contains no information about any math or physics studies related to the effects of CO_2 in the atmosphere. It contains descriptions of many negative effects of increases in atmospheric CO_2 and not a single mention of the beneficial effects. I believe these are glaring errors of omission that make all of the "scientific" conclusions in the book suspect. It is **consensus science** with little or no **hard science** to back it up or confirm the opinions.

Crista: I printed out and read the publication mentioned in your last email. It is very interesting and informative. However, I question some of the conclusions and explanations of the reasons for what is happening. I personally think that there are many other known and certainly some unknown causative factors that make for climate change, none of which seem to ever have been considered or studied and are certainly not even mentioned in that report. There is a list of some of these on page 51, and in a previous section of this blog.

You stated, **"Yes, there is some potential good to come from climate change, but on balance it will not be good, based upon everything that I have read and studied."** Have you ever read any reports from research into the positive effects of a warming planet? I doubt you have because the vast preponderance of research effort has been focused on the supposed detrimental effects. I know of no publication of the results of any research into the positive effects. In fact, other than the comment added to my blog by level head and a few of my own comments over the years mostly described in the section, "Bully for Global Warming" on page 34. I know of nothing positive mentioned or implied by anyone about global warming/climate change. That includes academia, the media, politicians, even those passionately against the whole AGW thing. Was your comment about potential good an abstract statement or have you read reports describing the possible

benefits? I would surely like to read them if there are any.

Another thing that troubles me. I mentioned to you the lack of any math or physics confirmation of the grossly misnamed *greenhouse* effect or use of the term *greenhouse* gasses in virtually all anthropogenic climate change reports including the one you mentioned. That is very difficult for me to accept or understand. The gas laws are firmly established science backed by math, physics and chemistry. The use of those laws to confirm the theory of AGW are essential at least to my acceptance of the theory. I would think that kind of confirmation would be paramount in the minds of all **real** scientists. Why the concerted effort to ignore or hide such calculations or dismiss them as irrelevant?

I am not arguing with you, but I do have the right to question when I do not see things as others do, even "experts" like you, who are sometimes in error. Incidentally, what education or training do you have in the fields of atmosphere, climate, weather, ocean circulation, or the gas laws and physics that deal with the broad field of atmospheric science? The history of science is liberally peppered with established theories that new insights have overturned. I also realize many of my explanations and ideas are suspect to the lettered **experts** because I have few letters after my name **(BS is all)** to indicate I know what I'm speaking of, but so what? The math and physics speak for themselves.

Why is it that none of the *lettered* and published experts will provide me with any kind of math or physics to prove the actual magnitude of the effect of increased CO_2 in the atmosphere?

Recent searches led me to another website with a commentary by Peter Kelemen, a professor of geochemistry at Columbia University's Department of Earth and Environmental Sciences. The article can be read at:

http://www.popularmechanics.com/science/earth/4338343.html.

In the article, Professor Keleman said, "It is equally essential to emphasize that alleged problems with a few scientists' behavior do not change the *consensus* understanding of human-induced, global climate change, which is a robust hypothesis based on well-established observations and inferences."

"*Consensus* understanding? Observations and inferences?" Does that now replace *hard* science proofs of the physical sciences? Do we now use "well-established observations and inferences" as the defining basis for science while avoiding math and physics? Direct me to the math, physics and chemistry proofs of AGW and I will become an ardent supporter. Until such time I will remain extremely skeptical especially since the math and physics calculations I have applied define a very different reality.

And why is their so little research into the beneficial effects of a warmer climate, or reports of benefits from any sources, regardless of cause? The work of my friend, Levelhead, is the sole example I have been able to find. There is a great deal of research into the negative effects. Where are all those studies balancing the negative? What about my friend's comments about what my essay didn't mention?

Go back to page 37 and reread what Levelhead, a Research Scientist at East Anglia has to say about the Benefits of Increased CO$_2$.

Oh yes, that report you recommended and that I downloaded is quite a testimonial to the environmental change and damage man has and is continuing to wreak on the planet. You will get no countering claim from me about those. My concern is that all this attention to a questionable cause moves us substantially away from directing research and corrective effort to the far more obvious problems cited in that report. **Actually, nowhere in the report did it mention the most obvious cause of the truly DEVASTATING human damage to the environment exacerbated by population overgrowth.** Of course, that little problem is unpopular because politicians have not found a way to turn it into cash in their pockets as they have global warming/climate change. The same could be said of the positive effects of increased atmospheric CO$_2$ and a warmer climate. Where are the studies of these? Surely there are some intelligent individuals, even scientists, in addition to Levelhead, who could mount well-based research into the positive effects. Why is there no such research or even suggestions?

Howard

Posted by HoJo at 2:10 P.M.

I never did get a response from her to this last email. Typical of "fundamentalist true believers," she refused to acknowledge or discuss any argument that refutes her position. It's like talking to a religious fundamentalist about evolution. They know they are right—end of discussion. As far as I am concerned, her silence gives testimony to the weakness of her arguments.

❖   ❖   ❖

More of the Benefits of Global Warming and a Warmer Climate

Wednesday, November 25, 2009 - this is the original post that started the exchange with the (real) scientist.

Climategate? - Warmergate? The truth about many of the scientists behind the global warming hoax has finally arrived. It's the Piltdown hoax (1912) of the 21st century—big deal. It's meaningless anyway because of two facts. First, There are at least a dozen factors known to affect climate more than atmospheric CO$_2$. Together, they make the CO$_2$ effect virtually meaningless. Second It's so easy to see that a warmer climate benefits all life on Earth that a six-year-old could understand it.

Another thing! Aren't the global warming fanatics looking at it backwards? Don't they have one very important fact upside down? Despite all that has been said on both sides, Global Warming, if it were real and its effects were substantial, it could be the best thing to happen to Earth in many millennia and infinitely more good than bad. If they knew this, would the doomsayers of global warming ever mention it? Of course not. It would certainly derail their gravy train.

In fact, global warming of any kind is infinitely beneficial to virtually all kinds of life on the planet. A simple proof: how much vegetation grows on Greenland now? How much vegetation (and all the fauna it supports) will grow there when the climate warms enough to melt all the ice. Think Michigan during the ice age with a mile of ice covering the entire state and look at it now. There is no arguing with those facts by anyone but an AGW ideologue

It is obvious that in the bigger picture and over the long range, global warming would be quite advantageous for life on earth, even humans. The benefits would far outweigh the losses. It would certainly be a lot better than another ice age. Of that there is no doubt. So I say, quit wringing your hands and crying of doom. Global warming could be one of the best things to happen since the last ice age ended.

So once more I say, burn that oil, burn that coal, pump out that CO_2! Let's heat up the planet and head for the beaches. How about palm trees in Labrador, oak forests on the north slope of Alaska or a seaside villa on the warm shores of Greenland? Sounds great to me.

While there will definitely be many winners, there are bound to be a few losers. Sorry, New York and LA, but you'll have to move a bit inland like quite a few other coastal habitats. I'm really sorry about Florida, but no change is ever all good. Besides, the lush semi tropical shores of Labrador will be habitat for former Florida flora and fauna while the lowlands of Georgia, the Carolinas, and the rest of the southern states all the way to the West Coast will be lush tropical paradises. That's certainly a much prettier picture than that painted by the doom sayers of the church of horrible global warming. I'd certainly like to hear a response from some of their members. Al, Eat your heart out!

The one glaring error the believers in detrimental global warming make is that it is detrimental. The fact is it would be almost entirely beneficial to mankind and to all life on the planet, and substantially so. Historically, the warm periods such as the *Medieval Climate Optimum (Medieval Warm Period)* were beneficial for civilization. Corresponding cooling events such as the **Little Ice Age** were uniformly bad news. Up to a point, the warmer our planet gets, the more life it supports and the more variety that life has. The cooler, the less life and the fewer species. All fossil records support that reality as far back as the Cambrian period, 500 million + years ago.

We have very good evidence that a warmer earth has far more benefits than negatives. For obvious reasons, global warming advocates never consider that very significant fact. It is well known from several sciences including botany, zoology, and paleontology that most all life has fared much better during warm periods on the planet than during colder periods. The amount and diversity of life were severely diminished during the ice ages, expanded as the glaciers receded, and were much greater during periods warmer than the present. As the borders between tropical weather and temperate as well as temperate and arctic on our planet move toward the polar regions, the number and variety of life increases as temperate areas gave way to tropical and arctic gave way to temperate. This is a well-established, logical fact, easy to understand.

Compare the number and variety of life as it now exists in the tropics, in the temperate zone, and in the arctic. Imagine what would happen if the tropics expanded and the arctic zone shrunk. Would that be beneficial to life or detrimental? Obviously it would be beneficial. I'll wager few people ever even considered that fact.

A case in point in human experience: The Vikings established a thriving community of farms, villages and churches in Greenland that lasted for at least 500 years during the *Medieval Warm Period* that was somewhat warmer than today. There is even evidence they established colonies in North America *(Vinland?)* during the same period. Then, when the *Little Ice Age* happened and arctic weather moved south over Greenland, they starved and disappeared leaving their fields, homes and churches where they stood. Greenland has for all practical purposes never been resettled.

Suppose the worst claimed by the gurus of global warming actually happened and sea levels rose a hundred feet. At the same time the tropics would expand 500 to 1,000 miles, and the temperate zones would move poleward about the same distance, Should this happen, huge areas of Canada and Siberia, among others, would become much more hospitable to life of many kinds—temperate forests and fertile farmlands. Since it is well known that warmer periods brought wetter weather patterns, many deserts would become lush and green. Talk about a green revolution—that would be a real one! Southern California, Arizona, New Mexico and West Texas valleys could become tropical jungles. The same could happen to the deserts: the Sahara, the Gobi, and even the Australian. Wouldn't that be a kick? Who knows, Greenland might even become green and habitable. One thing for sure, the Arctic ocean would become an open shipping lane. The overall benefit to life on the planet would be huge. So we would lose some land and a few cities along the coast and many islands would shrink or disappear, big deal! Since it would happen gradually, the people and much of their property could easily be moved elsewhere.

After realizing these facts, maybe Al Gore will write a new book titled, "A Beautiful Truth" that will tell the facts about global warming. Oh, but that wouldn't work with his liberal agenda, would it? Besides it could derail that highly lucrative cash cow the politicians of the world are riding to wealth and power. The new American politics trumps facts,

science, and all other forms of truth. Shades of the old Soviet Union.

Posted by HoJo at 3:04 P.M.

❖   ❖   ❖

Earth Hour - Is its Effort in the Right Direction?

Saturday, March 28, 2009

Earth Hour is a great idea except that most of the effort is aimed at an extremely minor problem that may be a non problem while ignoring the many major environmental problems plaguing us and especially the one overriding menace so very few ever mention.

I do not agree that there is overwhelming evidence that carbon dioxide produced by human use of fossil fuels is bringing about actual global warming that portends any great danger to humanity. Neither do I agree with those who deny any possible effect. The factual gulf between these two extreme opinions exists only in the minds of those that hold them. This is amplified by emotionally charged media frenzy and political fervor. In my opinion, *Global Warming,* that has recently been morphed into *Climate Change* by a cabal of financial and political benefactors of this questionable *fact* craze is, at its worst, a very minor problem. This is especially true when comparing it to many other very real problems and menaces facing humanity.

The almost spiritual global warming movement is gaining great numbers of ardent and vocal followers. Many of these are blind disciples who have absolutely no clue about the realities of climate change, the physics of the atmospheric *greenhouse* gases or whether there is even the possibility of many of the claims put forth by the high priests of global warming. This has been driven to virtually universal acceptance as an absolute fact because it serves the political, social, cultural and/or economic agendas of its proponents.

Perhaps it is a present day version of the Piltdown Man hoax foisted off on unsuspecting scientists and the public almost a hundred years ago in 1912. That hoax took forty years to be completely discredited. In 1923 Franz Weidenreich, an anatomist, reported that the skull was a modern human cranium and the jaw of an orangutan whose teeth were filed down. It took scientists thirty years to concede that he was correct. Like most of us, scientists hate to admit error on their part. Many of us cling to dogmatic positions long after an error is discovered and reality has become quite certain. Politicians and religious leaders are particularly so infected. History provides countless examples. Some were extremely damaging like the murder of Huss and the imprisonment of Galileo.

In spite of all this, there is one really good thing about the global warming movement. No matter how far it is from reality, it certainly has garnered the attention of the public, of the media, of governments, and of influential people. This brings attention to the overwhelming needs of our planet for serious concern, care, and attention. In spite of self-serving politicians and others who are in it for the money or power, some of the money and some of their efforts do have positive results. Fortunately, there are many dedicated people, mostly in the trenches, who are working tirelessly to prevent the destruction of our fragile environment. Tom Friedman writes about some of these people in his book, *Hot, Flat, and Crowded*. These people are at their best when they educate the public at all levels how protecting the wild environment makes sense economically as well as aesthetically. How a forest with all its interacting wildlife intact is so much more valuable long-term than the short-lived products made from the cutting of that forest. How ocean fisheries can produce protein food sustainably with proper management rather than the uncontrolled slaughter that has destroyed and continues to destroy a valuable but diminishing

resource. We can do to the earth what Easter Islanders did to their island home and so eradicate most life forms including us, or we can protect and sustain the valuable wild diversity on our small planet home.

Some practical information about the processes acting on our atmosphere

Here is some physical data no global warming proponent ever acknowledges. First of all, and most important, the term *greenhouse* as applied to atmospheric gases is a gross misnomer. In deference to its now common usage, I will use its new and very different meaning in this book. The actual process by which atmospheric gases retain heat energy and therefore cause the temperature of air to rise follows a very complex group of physical laws that are very different from what happens in an actual greenhouse. These laws involve the physical structure of the molecules of the various gases and how they resonate and/or rotate when they absorb infrared radiation or heat. Each molecule both absorbs and emits radiation at different rates for different wavelengths and at different temperatures, yielding varying amounts of absorbed, radiated and retained heat energy. The only way we can measure these effects is to do so collectively using a significant number of mixtures of various gases. These mixtures include a wide variety of those gases, water vapor among them. A glance at the data from one of the latest research studies on these phenomena reports, "Recent improvements in the spectroscopic data for water vapor have significantly increased the near-infrared absorption in models of the Earth's atmosphere." The full report is available at:

http://www.agua.org/crossref/2006/2005JD0 06796.shtml

Another report titled, *Water and Global Warming*, said, "Water is the main absorber of the sunlight in the atmosphere. The 13 million million (that's 13 trillion!) tons of water in the atmosphere (~0.33% by weight) is responsible for about 70% of all atmospheric absorption of radiation, mainly in the infrared region where water shows strong absorption. It contributes significantly to the *greenhouse* effect ensuring a warm habitable planet, but operates a negative feedback effect, due to cloud formation reflecting the sunlight away, to attenuate global warming. The water content of the atmosphere varies about 100-fold between the hot and humid tropics and the cold and dry polar ice deserts." The full article is available at:

http://www.lsbu.ac.uk/water/vibrat.html

❖   ❖   ❖

There is another article on the effects of CO_2 at:

http://brneurosci.org/.htm

Any global warming from the effects of CO_2, if indeed it exists or poses any danger at all, is grossly distorted relative to the facts at hand. Most of the data used to show global warming are at best statistical and at worst, consensus and anecdotal. Both types of this nebulous data provide great opportunities for opinions (and agendas) to affect the resulting conclusions. We know for certain that addition to the atmosphere of any gas will contribute that gas's infrared absorption and radiation properties with all their complexities. Actually, all gases in the atmosphere have some *greenhouse* effect. This includes, nitrogen (75.0% - 78.08%), oxygen (20.11% - 20.95%), argon (0.89% - 0.93%), and carbon dioxide (0.035% - 0.038%). The percentages in parentheses are of air at sea level. Ranges are shown because air also contains a variable amount of water vapor (from 1- 4% $\pm$ 0.25%) and trace amounts of other gases. Each gas has a complex rate of infrared absorption, transmission, and emission at various infrared frequencies. Atmospheric water vapor is from 25 to 120 times the amount of CO_2 in the atmosphere and has about 25 times the net temperature effect of the same amount of CO_2 depending on various conditions. Bear in mind that net effect is the difference between energy from the Sun coming in that heats the atmosphere and energy from the atmosphere going out into space that cools it.

Energy in is radiant energy from the sun being absorbed by atmospheric gasses and Earth surface materials. It includes that convected from surface materials of the earth into the atmosphere. Energy out is that radiated from the surface that freely radiates into space and is not absorbed by those gasses on its way out through the atmosphere as well as energy emitted by atmospheric gases out into space. Considering the varying amounts of each gas in the atmosphere, results in a range for heat retention of water vapor are between 500 and 3,000 times that of CO_2. This number varies with temperature, altitude, location and water vapor content. All told it is an extremely complex system with many variables. If all factors are considered in their proper proportions and even if the amount of CO_2 doubled, it would have a negligible effect on average global temperature.

The warmer air becomes, the more water vapor it can hold. Remember the weatherman's favorite "dew point" predictions? When the temperature lowers to that point, the air can hold no more water vapor so it condenses out as dew or rain in the big picture. Using the same rationale as the global warming folks use for CO_2, increasing amounts of water vapor would cause a much larger increase in atmospheric temperatures than CO_2 resulting in still warmer air and still more water vapor. Shouldn't this lead to a runaway *greenhouse* effect? Wouldn't this drive atmospheric temperatures higher and higher until the oceans boil and all life is extinguished? Obviously this has not happened so something about these assumptions must be wrong for water vapor and CO_2 as well.

Water vapor adds another major factor to the mix. That is the heat of vaporization or condensation of water. A tremendous amount of the sun's radiant energy evaporates water all over the earth. All of that energy enters the atmosphere in water vapor. The warmer the ocean or land, the more energy goes into evaporating water into the air. When all this water vapor condenses out as rain, that energy is released and the air warms. This is the driving energy that causes the air to move and creates windstorms, tornados and hurricanes. For all practical purposes, the CO_2 content of the air has zero effect on the amount of energy that goes into the atmosphere or heats the air when water condenses.

One huge factor that man has affected greatly is the water vapor that green plants give off and particularly dense rain forests. Our continuing decimation of all types of rain forests is removing a huge source of water vapor that formerly entered the air. One example of this effect was used incorrectly as an example of global warming, which it was not. The disappearing snows of Mount Kilimanjaro are not an effect of global warming. Studies have shown that the cutting of the forest around the base of the mountain substantially reduced the amount of water vapor in the air flowing up the mountain. The result was that both the rainfall and snowfall on the higher slopes have been reduced dramatically. This is one correct example of where human activity has interfered with nature. Deforestation worldwide has done far more damage to our environment and effected climate far more than even tripling the amount of CO_2 in the atmosphere could do. It alone could arguably be responsible for any temperature increase over the last hundred years as a reduction in the amount of water vapor would reduce cloud cover resulting in less of the sun's energy reflected away. Why don't we do something about that?

Whatever the effect of carbon dioxide, it is so small in comparison as to raise questions about the real amount of the danger it poses. Certainly it is not the degree of danger claimed by the high priests of global warming. I seriously question the validity of the often quoted phrase, "Overwhelming numbers of scientists support the theory that man's use of fossil fuels is bringing about catastrophic global warming." In the first place, the worldwide destructive clearing and burning of rain forest results in putting far more net carbon dioxide into the air than all the vehicles in the entire world. Second, shrinking rain forests mean

less water vapor is released into the air. This could in turn mean less rain and snow where the air over land is drier. The questions remain, does the evaporation from the oceans increase and make up for this loss, and what effect does the drier air have on cloud cover and the resulting reflection of the sun's energy away from the earth? All of these interacting variables have much larger net effects on global temperatures than CO_2. Because of this, *Overwhelming numbers of scientists* may have no real clue about the degree of effect that CO_2 might have on atmospheric temperatures leading to global warming. Obviously it is much smaller than that of water vapor.

To explain why I make this statement, I have quoted biologist Edward O. Wilson. A Pulitzer prize winner, Wilson is an experienced and admired biologist and author of numerous books. He may have given us a clue as to why *many scientists* may not be good judges of climate change and of global warming or how dangerous it is. In his book, **The Creation, an Appeal to Save Life on Earth**, he speaks about *scientists*, who they really are, and why the *scientific method* works. The following is a revealing quote from the eleventh chapter, **Biology Is the Study of Nature**. The italics in the quote are my comments.

❖   ❖   ❖

The quoted section from E. O. Wilson begins:

I will offer now an account of the concept and practice of science and in particular biology, the discipline most immediately relevant to human concerns.

I hasten to add I do not mean scientists. Most researchers, including Nobel laureates, are narrow journeymen, with no more interest in the human condition than the usual run of laymen. Scientists are to science what masons are to cathedrals. Catch any one of them outside the workplace, and you would likely find someone leading an ordinary life preoccupied with quotidian tasks and pedestrian thought. Scientists seldom make leaps of imagination. Most, in fact, never truly have an original idea. Instead, they snuffle their way through masses of data and hypotheses *(the latter are educated guesses to be tested - global warming?)*, sometimes excited, but most of the time tranquil and easily distracted by corridor gossip and other entertainments. They have to be that way. The successful scientist thinks like a poet, and then only in rare moments of inspiration, if ever, and works like a bookkeeper the rest of the time. It is very hard to have an original thought. So for most of his career, the scientist is satisfied to enter the figures and balance the books.

Scientists are also like prospectors. Original discoveries are the gold and silver of their trade. If important, they can buy collegial prestige, and with it wider fame, royalties, and academic tenure. Scientists by and large are too modest to be prophets, too easily bored to be philosophers, and too trusting to be politicians. Lacking in street smarts, they are also easily fooled by confidence artists and sleight-of-hand tricksters *(and global warming promoters?)*. Never ask a scientist to test the claims of paranormal phenomena. Ask a professional magician.

The power of science comes not from scientists but from its method. The power, and the beauty too, of the scientific method is its simplicity. It can be understood by anyone, and practiced with a modest amount of training. Its stature arises from its cumulative nature. It is the product of hundreds of thousands of specialists united by one commonality of the scientific method. Few scientists know more than a small fraction of available scientific knowledge, even within their own disciplines. But no matter: their fellow scientists are continuously testing and adding other parts, and the entire body of scientific knowledge is easily available. The invention of this remarkable engine of testable

learning was the one advance in human history that can be called a true quantum leap. But it attained its preeminence relatively late in the geological life span of humanity, and only after human intellect had traveled a long, tortuous path dominated by tribalism and animated by religion.

Let's try to establish a rough chronology. Millions of years ago there were only animal instincts. Then, probably at the man-ape level, the rudiments of materials culture were added. With still higher intelligence there followed a sense of the supernatural, whereupon demons, ancestral ghosts, and divine spirits peopled the human mind. Without science there had to be religion, in order to explain man's place in the universe. Born of dreams, its images were enshrined in the culture by shamans and priests. The gods made man. Those that lived in surrounding Nature gave way to gods of sacred mountains, in distant places, and in the heavens. Somewhere and somehow back in time, these divine humanoids had created the world, and now they governed man. Humans in their evolving self-image, rose above Nature to follow the gods as children and servants. Tribes led unwaveringly by their personal gods were united and strong. They defeated competing tribes and their false gods. They also subdued Nature, erasing most of it in the process. Their destiny, they believed, was not of this world. They thought of themselves as immortal, no less than demigods.

Along the way, commencing in Europe in the seventeenth century, a radical alternative self-image emerged. Art and philosophy began to disentangle themselves from the gods, and science learned to operate with full independence. Step by step, often opposed by the followers of Holy Scripture, science constructed an alternative world view based on a testable and self-reliant human image. Doubling in growth every fifteen years during most of the past three and a half centuries, it has looked into the heart of living Nature, finding there a previously vast and autonomous creative force. This image has subsumed religious rivalries and reduced them to intertribal conflict. Science has become the most democratic of all human endeavors. It is neither religion nor ideology. It makes no claims beyond what can be sensed in the real world. It generates knowledge in the most productive and unifying manner contrived in history, and it served humanity without obeisance to any particular tribal deity.

—*Edward O. Wilson*

End of quote.

There is one certainty about the global warming movement. It has become a *cause celeb* among the elite and generated huge amounts of cash, mainly for politicians. This money is in the form of numerous, varied, punishing new taxes, cap and trade agreements, and expensive regulations. These taxes and the hundreds of global warming organizations constantly soliciting donations have turned it into a huge, self-perpetuating cash cow for its promoters and benefactors. This will guarantee its continuation long after it is proven untrue or at best, overblown far beyond any real danger.

The menaces that are a far greater danger to life on Earth than global warming at its worst.

In this writer's opinion there are numerous other, far more dangerous menaces facing humanity than global warming in its worst case scenario, anthropogenic or no. I will briefly mention one of those, population. Considering our exponentially expanding population and our steadily diminishing resources one would think concern for this would be paramount in the minds of all thinking people. This is certainly the most serious and overriding one of several score of serious menaces we are facing right now. It alone is the driving force of many of our problems and especially those related to the environment and food supplies. Will we continue concentrating our attention on things like global warming and the next American idol or soccer champion while major problems fester and grow with little comment or attention? Like Nero, humanity fiddles as the world burns.

The growing shortages and rising prices of food are bringing attention to the fact that something is going drastically wrong. Unfortunately, most reports condemn those involved in the food industry they see as responsible for rising prices. They make no mention of population growth, the very real reason for the shortages bringing about rising prices. The same could be said for many other of our rapidly disappearing resources. It seems politicians and the media are far more interested in using invented menaces as tools to promote their own agendas to the masses rather than in finding solutions to real and dangerous ones like overpopulation of humans.

"We have been Godlike in our planned breeding of our domesticated plants and animals, but we have been rabbit-like in our unplanned breeding of ourselves."

—*Arnold Toynbee*

This Real and Present Danger Is a Far Greater Threat to Life on Earth than Global Warming.

Some time ago I went to my family physician for pains in my knees and back. After the examination and his recommendations I asked a simple question, "Doc! What's happening to me?" His simple, straightforward answer said it all with great accuracy, "You're wearing out." Let me say that is exactly what the human species is doing to our world, we're wearing it out and far beyond its ability to heal or repair itself. Deterioration is accelerating and will continue to do so until and unless something stops the insanity that is population growth. Nothing else will work! Nothing! We can cry all we want about disappearing species and growing extinctions, but the fact is simply that human reproductive success and over achievement is leading inexorably to the obliteration of all competing species, and much quicker than we think. Look at what has happened in the last one hundred years as we became more efficient at catching wild food and destroying wild habitat. Life on the earth will not handle another hundred years like the last. The greatest extinction of species since the Permian is not over. It is just beginning!

Tragically, we do virtually nothing about it as it expands geometrically. Our nearly total denial of the menace of overpopulation is not only grossly foolish, but it makes our efforts to curb global warming at its worst look like child's play. The worldwide forces of cultures, religions, and raw animal instincts that are bringing billions of excess humans into a world with limited resources are a certain doomsday juggernaut that may already have passed the point of no return—the point of irreparable damage. We are in the midst of the greatest extinction of species since the Permian. The great sink of wildlife in Africa is fast disappearing down growing millions of hungry throats as "bush meat." Still, the human population there continues to explode, even in the face of starvation, abject poverty, and disease.

Many once limitless ocean fisheries are rapidly disappearing. The Atlantic cod and salmon fisheries are gone. The Pacific salmon are also going. Alaskan waters are now fished so efficiently and relentlessly this source of protein could be decimated within a decade. Wild food from the ocean could be a thing of the past within a relatively short time. Fish and sea creatures eaten by fish and other sea creatures are part of a cycle that maintains a balance of sea life. Seafood eaten by man causes an imbalance because all of that protein is removed from the oceans completely and forever. Every ounce of protein we remove from the ocean depletes that source and once gone, it will not return.

Corn and soybeans diverted from the food chain for alternate fuels are causing food prices to rise. This is causing great anger in places like Egypt where the price of wheat imported mostly from the US has nearly tripled recently. In other parts of the world, diverse food crops are being replaced by single money crops. Generally forest clearers use slash-and-burn techniques to clear land, but on a much larger scale than traditional practices. Instead of burning a mere 2-10 acres, agriculturalists burn hundreds to thousands of acres after felling a tract of forest and leaving it to dry. Burning releases nutrients locked up in vegetation and produces a layer of nutrient-rich material above the otherwise poor soil. The cleared area is quickly planted and supports vigorous growth for a few years, after which the nutrient stock is depleted and large amounts of fertilizer are required to keep the operation viable. In many tropical areas, diverse rain forest is being destroyed to grow oil palms to feed the demand for biodiesel. Everywhere, natural biodiversity is being replaced by row upon row of single plant crops. This results in major loss of diverse plant ecosystems and habitat for countless wild creatures. When the land is suitable for agriculture, generally large single cash crops like rice, citrus fruits, oil palms, coffee, coca, opium, tea, soybeans, cacao, rubber, and bananas are cultivated. Add to this the loss of crop land to urbanization, salinization of irrigated lands, soil depletion, and desertification and the view for the future is not pretty. The worst case scenario of *greenhouse* gases pales in comparison with the results of the best and most conservative estimates of population growth. Also, when crunch time comes it will be a sudden catastrophic collapse, not a slow change. Hell will probably be a fair place in comparison.

Availability of just one simple device could greatly reduce the use of wood for cooking in many parts of the world. A solar stove using a concave mirror to focus sunlight in such a way as to cook food would save tons of wood fuel when and where sunlight is available. I had such a stove years ago. It was small, inexpensive, and worked very well on sunny days when we were camping. It boiled water, cooked vegetables, and broiled meat to perfection. We even used it to bake biscuits. It folded up into the size of a small, thin briefcase. Made of aluminum, it was light and easy to carry. This is merely one small item of the thousands of small, effective solutions that could be put to use to help third world people.

❖   ❖   ❖

More Concerns About Global Cooling

Climate is an extremely complex system that we have been studying for a long time up to and including the age of the supercomputer and computer modeling. Still, we have hardly touched the surface as can be attested to by the accuracy of our current weather forecasting. For example, in spite of all our technology, predictions of the frequency, location, and path of any hurricane are fraught with pure conjecture. We can't even hope to predict the intensity of any hurricane season. Witness the 2006 season. It was predicted to be one of the worst ever. Instead it turned out to be one of the mildest, the opposite of the predictions of some of our weather scientists and their supercomputer modeling. The hurricane prediction record of the years since 2006 has been much the same. How about the accuracy of your local weather forecaster? How often does he or she miss the mark predicting a day ahead? And remember, global cooling is also climate change.

The world's climate system is infinitely more complex than a single hurricane season. It moves in cycles and eddies that run from seconds to millennia. During the early 1960s, climate pundits feared we were heading into global cooling and needed to prepare for a drier, cooler time with lower sea levels.

According to many scientific studies of past frigid periods we are past due for the onset of the next ice age. Hubert Lamb of the UK Met Office dominated the 1961 UN meeting on global cooling. A founder of the Climate Research Unit at East Anglia, he was one of the world's top climate scientists. He warned that people had become complacent about climate at a time when population growth, cold, and drought could seriously damage their food supplies. (The Norse in Greenland perished of starvation after five hundred successful years when the *Little Ice Age* destroyed their crops.) In historic times the climate has veered from warmer than the present, the *Medieval Warm Period*, to the much colder conditions of the *Little Ice Age* from which we may still be emerging. Evidence shows that much of the Sahara and the Middle East held lush vegetation and crop land ten thousand or so years ago while northern Europe and America were covered with up to a mile of ice.

Many scientists and climatologists have been predicting the onset of a new ice age. Based on past climate cycles from warm ages to ice ages and looking at the major factors that influence how much energy the Earth receives from the sun, the most likely scenario is change to a much colder, ice-age climate, and soon. Anecdotal evidence of climate change that is far more damaging than global warming is being considered by climatologists who are not overwhelmed by the global warming crowd. Indeed, there are several very real happenings that do not support global warming. Many are anecdotal, but the overwhelming evidence paints a very different picture than the one touted by global warming proponents.

I recently visited Alaska and spent a day in Glacier Bay. While there I learned some interesting facts, mostly from a recent publication about the glaciers. Since the mid 1700s, Alaskan glaciers have been known to be steadily receding. Early explorers found glacier ice all the way to the mouth of what we now call Glacier Bay. There were maps in the book with lines showing the dates of glacier terminuses from the 1700s to 2007. All the glaciers were shown to have steadily receded until the early 1990's. Since that time all of these glaciers have advanced steadily. In recent years, average global temperatures have dropped. I also learned that arctic sea ice has been increasing rapidly since 2004. Recent tests show arctic ice to be thicker than it has been for many years. I wonder why the media has not made the public aware of these facts? Sure, this is anecdotal, but so is the earlier information about melting arctic ice.

Over the two winters (2008-2010), anecdotal evidence for a cooling planet has exploded. China has its coldest winter in 100 years. Baghdad sees its first snow in all recorded history. North America has the most snow cover in 50 years, with places like Wisconsin the highest since record-keeping began. Record levels of Arctic and Antarctic sea ice, record cold in Minnesota, Texas, Florida, Mexico, Australia, Iran, Greece, South Africa, Greenland, Argentina, Chile -- the list goes on and on. Already this winter, December 2008, has seen some of the coldest and most severe winter weather ever recorded.

Added May 2013: the past winter has been the coldest on record for most of the northern hemisphere, even colder and more severe than 2008-9. In Scandinavia, the onset of spring and the blossoming of fruit trees and spring flowers was at least three weeks later than usual. May was unusually cold in all the countries usually warmed by the gulf stream.

It is interesting to note here that the onsets of previous climate changes have not been gradual, but quite sudden in relative terms. The native people who live near the mouth of Glacier Bay in Alaska once lived several hundred miles north of their present home. At about the same time, the Norse in Greenland were being wiped out by crop failures from the onset of the *Little Ice Age*. These Alaskan native legends describe an advancing glacier moving south "as fast as a running dog." There are tree stumps and other evidence exposed when the glaciers receded to their current terminus showing the glaciers once had receded far beyond their current position during the *Medieval Warm Period* that ended about a millennium ago.

No more than anecdotal evidence to be sure, but now that evidence has been supplanted by hard scientific fact. All four major global temperature tracking outlets (Hadley, NASA's GISS, UAH, RSS) have released updated data. All show that over the past few years, global temperatures have dropped precipitously.

Meteorologist Anthony Watts compiled the results of all the sources. The total amount of cooling ranges from 0.65C up to 0.75C -- a value large enough to erase nearly all the global warming recorded over the past 100 years, all in one year's time. For all sources, it's the single fastest temperature change ever recorded, either up or down.

Scientists quoted in a past DailyTech article link the cooling to reduced solar activity which they claim is a much larger driver of climate change than man-made *greenhouse* gases. The dramatic cooling seen in 12 months time seems to bear that out. While the data doesn't itself disprove that carbon dioxide is acting to warm the planet, it does show clearly that more powerful factors could now be cooling it.

Let's hope those factors stop fast. Cold is more damaging than heat. The mean temperature of the planet is about 54 degrees. Humans—and most of the crops and animals we depend on—prefer a temperature closer to 70.

Historically, the warm periods such as the *Medieval Climate Optimum or Warm Period* were beneficial for civilization. Corresponding cooling events such as the *Little Ice Age* though, were uniformly bad news.

The truth of the matter is that we are affecting the climate by adding CO_2 to the atmosphere. That's roughly the same kind of truth as the fact that pouring a bucket of water into lake Erie will raise the lake level. That is about the same order of rise that can be attributed to increased CO_2 in the atmosphere. We have very little definitive knowledge of how much effect raising or even doubling the amount of CO_2 in the atmosphere will have in the long run. Other than using the calculations noted earlier in this article, we certainly are unable to hazard more than an intelligent guess as to what or to what degree the effect might be relative to other factors. The following are known to affect climate and average world temperatures as much or more than any increases in atmospheric carbon dioxide.

These Are the Ten Processes:

1. The wobble of the earth's axis increases or decreases the retention of energy from the sun. (22,000-year cycle)

2. The eccentricity of the earth's orbit increases or decreases the energy we receive from the sun. (12,000-year cycle)

3. The variation of energy output by the sun. (1,400-year cycle)

4. Variations in snow cover—snow reflects heat.

5. Variations in cloud cover—clouds reflect heat.

6. The variation in cosmic rays causes a variation in cloud cover.

   (**IMPORTANT:** See, A Stellar Revision of the Story of Life, page 16.

7. Dust and sulphate in the air can absorb or reflect heat.

8. Ocean temperatures and circulation.

9. Volcanic activity (For instance, the eruption of Mount Pinatubo in the Philippines brought on several years of cooler temperatures, probably from the volcanic dust it placed in the stratosphere that reflected infra red from the sun out into space.)

10. Winds—as winds increase, dust from dry farmland and deserts, enters the air. (Gobi desert dust sometimes reaches as far as our west coast. Saharan dust frequently covers the Amazon basin.)

It is also important to know that items one through three above produce many complex variables with many secondary effects on the temperature of the earth. For example: they all affect the power of the "solar wind." This powerful force affects our planet strongly and varies widely on an almost hourly basis. The solar wind is a stream of charged particles or plasma ejected from the upper atmosphere of the sun. It varies widely in its power and occurrence sometimes in very short periods of time—days or even hours. Though the earth is protected from direct exposure to this energy by its magnetic field, some of this energy does reach the surface. The effects it has on our atmosphere and climate are unknown. The noticeable effects include auroras which are relatively harmless and magnetic or EMF disturbances which can have devastating effects on electronic equipment including computers, communication equipment, and navigation systems. These solar "storms" have even disrupted electric transmission shutting down large sections of the power grid. These forces can also strip away portions of our atmosphere forming a "tail" pointing away from the sun in much the same manner as a comet's tail. It is thought that Mars once had water and an atmosphere similar to earth's, but it was mostly stripped away by solar wind over millions of years. Though the actual effect on the temperature of earth is unknown, the solar wind could be a major and irregular modifier of atmospheric energy and thus climate.

The rest of the factors are also varying and can be interdependent yielding an extremely complex mix of variables needed to produce any valid computer simulations. The effects of changes in the amount of CO_2 in the atmosphere on global temperature are probably smaller than those of any of the other factors. Records from ancient ice cores that show the CO_2 content of air to be much smaller during previous ice ages. Global warming proponents say this is proof that lower CO_2 levels cause lower air temperatures and higher CO_2 levels bring about higher air temperatures. Actually it is far more likely that temperature variations are the cause of changes in CO_2 levels rather than the result. This is evidenced by the fact that the ups and downs of CO_2 levels **follow** those of temperatures. Effects always follow behind causes, they never precede them.

So you folks who purchased all those carbon credits and donated all that money to various anthropogenic global warming cause organizations can kiss that money good bye. I'm certain it is long gone with nothing to show for its passing. Perhaps when the ice sheets begin advancing you can join a new "global cooling" movement and pay for carbon debits to help warm the planet.

An Unusual Conclusion from the research conducted during the writing of my book, *Energy, Convenient Solutions*:

- Almost Free Energy!

After all the research of information about energy and fuels used to write my book, *Energy, Convenient Solutions*, it became my conclusion that there are one or two best possible solutions for practical, affordable energy and its use. In addition to the huge economic benefits to our nation, these systems could answer all the real or imagined concerns about CO_2 caused global warming. I would urge those in power to consider doing everything required to make such systems realities. The total of both systems involves electric vehicles, EV's and plug in hybrid electric vehicles, PHEV's powered mostly by electricity from batteries charged from an electric grid supplied by geothermal or wave action power.

GEOENERGY: is the most abundant and widest spread source of energy on the planet, yet it is rarely addressed. It is virtually inexhaustible, economically available, non polluting, non carbon dioxide emitting, and grossly underutilized. At present a geothermal power plant costs about the same as a coal-fired plant of the same capacity and has a smaller footprint. Once built, no fuel system is required so maintenance is the only ongoing cost. Geothermal technology is now in its infancy. Improved technology of the heat transfer from deep in the earth to steam generators on the surface could lower the cost of power to a fraction of other systems. It is potentially the least costly form of power generation available and certainly has the lowest environmental impact. It is an environmentalist's dream come true. Its use requires drilling for heat almost exactly like drilling for oil, a well-developed technology. Why so few people ever even mention it is a mystery. The development of geothermal energy to replace retiring coal plants and provide the necessary increase in electric generating capacity could be the best way for our future. With technology that is presently beginning to grow, improvements in cost and performance could easily make it the best and most economical domestic source of electric power. This would satisfy complaints of both the global warming and anti nuclear crowd at least as far as generation of electric power is concerned. On the world stage, GEOENERGY is practical in most areas of the globe. It is especially available in Africa and could be a major factor in curing that continent's serious ills.

There is one type of system that could be engineered and built with today's technology. Using oil-well drilling technology, drill one or more wells down to the level of rock hot enough to be used efficiently, more than 600°F. Case each well and seal the bottom. Insert a smaller pipe inside the casing. Pump liquid sodium down the inner pipe and up the outer one. Run the hot liquid sodium through a heat exchanger to heat water to provide high pressure steam to drive steam turbines. This system is currently being used in nuclear power plants and is a well-established technology. The geothermal plant steam section would be virtually identical with the steam section of nuclear power plants, but without the thermonuclear heat source.

The difference would be that there would be none of the following costs or dangers: radioactive fuels and waste, fuel requirements of any kind, mining or shipment of fuel, disposal of radioactive waste, and/or carbon dioxide emission. This is an environmentalist's dream come true. All we need do is design and build the plant using existing technology. Once built and activated, the only costs would be maintenance and operation of the plant. Why hasn't this been done yet? Hmmmm?

Wave Action Energy: One of the newest concepts is to derive energy from the sea. Wave action generators of several types are being studied and developed. Pilot operations are being constructed in several places around the world. These generators convert the energy of sea waves into electrical energy by several methods. Buoys tethered to the sea floor move up and down in an elliptical path, turning a generator. Others simply use that elliptical motion and convert it into electricity. In some installations, waves are funneled up the shore through a type of valve that directs the water through turbine generators by the force of gravity as it returns to the sea. Most of these technologies are quite new and unproven, but a tremendous amount of energy is available in the wave motions of the sea

There is an interesting article about an efficient type of wave generator in the July 2009 issue of Smithsonian Magazine. Developed by Professor Annette von Jouanne of the university of Oregon, the generator uses magnets attached to a float that slides up and down a relatively fixed stator containing a coil of wire. The motion of the magnet passing the coil generates electricity like it does in any generator.

The article can also be viewed on the Internet at

http://www.smithsonianmag.com/science-nature/Catching-a-Wave.html.

Practical use of this readily available energy store could be another valuable source of cheap electricity. This is now being touted by some as the best possible source of new power with virtually unlimited potential. I do not see it as having any advantage over geothermal at this time.

Posted by HoJo at 10:37 A.M.

❖   ❖   ❖

NASA's Global CO$_2$ Surveyor from Feb 9, 2009 Aviation Week

Saturday, March 21, 2009

My son recently sent me a copy of an article in Aviation Week about NASA's global CO$_2$ surveyor satellite that unfortunately didn't achieve orbit and was destroyed. I could not find any links to the article, so I attached it to the email sent to most of my family and friends. In the article I found some interesting statements that prompted the following response:

Dear Mike:

I often become suspicious of the veracity and the political motives of articles published by some "scientists" when those articles contain statements that are patently false or at least misleading. I also view as suspect the common assumption made that CO$_2$ generated by human activity is a major factor in and causes catastrophic global warming. This one assumption runs counter to conclusions from many long studied and well-known factors that affect climate. The following paragraph is from that article. It caught my eye because one little sentence in italics, the basis for the accuracy of the entire project, is untrue.

"Carbon dioxide molecules aren't measured directly; the instrument tabulates the absorption of sunlight by CO$_2$ and molecular oxygen molecules before and after sunlight is reflected off the Earth's surface. Since each molecule has a unique infrared signature, they can be singled out and counted. There are two detectors for CO$_2$ because it is easier to spot near the Earth's surface at 1/61 micron and in the atmosphere at 2.06 microns. *The molecular oxygen A-band channel acts as a survey control because its presence in the atmosphere is constant.*"

In rebuttal I offer a substantiated quote from my book, ***Energy, Convenient Solutions***, and an article

I published in 2007. The full article on global warming can be found at http://gulfstream.blogspot.com. Should you want to read the section in my book it is available on the web at http://ecsexcerpts.blogspot.com.

From page 169 of the book: "With the exception of hydrogen, all gaseous, liquid, and solid fuels produce carbon dioxide when burned in any energy process. In addition, the production of hydrogen by any means other than by electrolysis, using energy from nuclear, wind, water or tidal power plants will add carbon dioxide to the atmosphere from both the energy and the raw materials used to create the hydrogen (coal-fired power plants for instance). It is interesting to note that for each pound of carbon oxidized to carbon dioxide, four pounds of oxygen are removed from the atmosphere. For every thousand tons of CO$_2$ added to the atmosphere, eight hundred tons of oxygen are removed. In all the concern about CO$_2$ there has never been a single mention of that fact."

The obvious conclusion is that the amount of oxygen in the atmosphere **is not a constant** as stated in the article in Aviation Week.

It should be interesting to note that I do not agree with either those who say increasing atmospheric CO$_2$ has a catastrophic effect on climate, or with those who argue it has absolutely no effect. It is just that the actual size of that effect, though real, is so small as to be insignificant compared with other, well-known factors. *See page 51.* In the same way, pouring a bucket of water into a small lake raises the lake level. I do not think shore dwellers need to fear that will flood their homes. This puts me solidly against global warming proponents in government who use it to gain power and as a huge cash cow of tax revenue. In addition there are literally thousands of groups using proven tactics to frighten the public into compliance, donations, and mindless support.

❖   ❖   ❖

Anthropogenic Global Warming, Fact or Profitable Fraud and Hoax?

Thursday, November 08, 2007

There are as many opinions about global warming and man's part in it as there are politicians, climatologists, weather scientists, media people, even celebrities and bloggers. None of these sources are particularly knowledgeable about the complexities of weather. There is a great paucity of good solid scientific facts. Everyone with a political ax to grind, a grant to go after, readers or watchers to excite, an idea or position to sell, or who have anger at something or someone, is jumping on the global warming bandwagon.

There are a few indisputable facts—or as indisputable as any "facts" can be. Remember, "One man's fact is another man's opinion and still another's fantasy."

That is because most facts are trumped by beliefs, a very human trait. Consider this quote from a paper I wrote many years ago.

"Perception, ah yes, perception, it is what drives our decisions, controls our emotions of love, anger, joy, disappointment, friendship, hatred, virtually everything we think or react to. Perception overrules facts, logic, and reality. Whether from love, avarice, or foolishness, and no matter how removed perception is from truth, it still rules us and determines our life decisions. We do not live in a real world, but live totally in a world created by and subject to our perceptions."

—*Howard Johnson, 1960*

❖   ❖   ❖

Most facts are trumped by perceptions or beliefs. That is a very human trait. These "facts" in consideration include:

1. Increasing the percentage of CO_2 in the air causes the atmosphere to hold more heat. CO_2 is such a small portion of air, its effect is infinitesimal.

2. Increasing the percentage of moisture in the atmosphere causes the atmosphere to hold more heat. Water controls numerous orders more atmospheric heat than all other greenhouse gasses combined. It also varies greatly from place to place and over time.

3. Increasing the percentage of methane in the atmosphere causes the atmosphere to hold more heat. It has a much larger heat capacity than CO_2.

4. Increasing the percentage of any of the *greenhouse* gases in the atmosphere causes the atmosphere to hold more heat

5. Pouring a bucket of water into the ocean adds to the level of the oceans. How much depends on the size of the bucket and how many buckets are used.

6. Burning any fossil fuel adds carbon dioxide to the atmosphere.

7. Animal life on land or in the seas adds carbon dioxide to the atmosphere and removes oxygen.

8. Green plant life on land or in the seas removes carbon dioxide from the atmosphere and adds oxygen.

9. Variation in the energy output of the sun affects global temperatures.

10. Variation in the circulation of the oceans affects the distribution of heat in the atmosphere.

11. Variation in galactic cosmic rays affects the amount of cloud cover over mostly tropic ocean water.

12. Variation of cloud cover affects global temperatures as clouds reflect the sun's heat energy.

13. There are dozens of demonstrable natural systems that affect global temperatures as much or more than does the *greenhouse* effect of CO_2. (The cosmic ray effect on cloud cover is one)

14. Most of the period between 800 CE and 1300 CE was much warmer than the current century.

15. These are but a few of the countless facts that effect climate and global temperatures.

16. Anecdotal reports of unusual local temperature and moisture conditions are completely useless as an indicator of global warming because of the huge scale of size and time differences between those reports and the global scales of climate changes over millennia.

17. In the 1970s after a series of very cold winters, a significant number of climatologists and other scientists were discussing global cooling and the coming ice age. The media and others reported this extensively to the great concern of the public. Since the winter 0f 2007-08 and its extreme cold and hordes of snowstorms, we're hearing the same kinds of rumbles. This is of course, overwhelmed by the constant propaganda of the AGW crowd.

Maybe Chicken Little was right.

All of this is intensified by our fascination with sound bites and short reports and lack of interest in lengthy, serious considerations of those same subjects, especially biodiversity. Most people get all of their news from the highly biased main stream media, NBC, CBS, ABC, and CNN. They often repeat verbatim, the biased, highly questionable and even false information from these far left sources. They simply lack the will or the time to do any in-depth reading or study about this very complex subject. Those who do such reading and study, and even research the subject have widely spread differences of opinion. This is the reason for such a controversy. Into this confusing mix come politicians and "expert" celebrities with huge agendas of their own who capitalize on this by arousing fear and other emotions in the all too gullible public.

With all this dust and smoke obscuring reality, even for those qualified to have a significant opinion, it is no wonder the fire is very hard to find if indeed it even exists. All this *smoke and mirrors* make for much confusion and those who live by emotional appeals make much use of the resulting disarray.

Here are two of the thousands of articles available from the Internet on the subject. They seem a bit more rational than many. Most are emotional diatribes driven by anger or a political agenda.

http://www.heartland.org/Article.cfm?artId=22218

http://petesplace-peter.blogspot.com/2007/06/weather-channel-backs-global-warming.html.

❖　　❖　　❖

email to Bob Grimm about James Hansen's Global Warming Article

Thursday, April 26, 2007

Bob:

I read with interest the words of James Hansen and I do appreciate you sending them to me. It reminds me of an old philosophy of action called, loosely, "tool theory." It's been around for a very long time, probably several centuries. Roughly it states that man (that includes both male and female) will try to use the tools he is mentally equipped and trained to use in efforts to solve any problem, make a chair for example. A carpenter will use hammers, saws, chisels, wood and nails to solve his problem and build the chair. A plumber, on the other hand, will try to use pipe, fittings, pipe wrenches etc. to build a chair. A stonemason might use stone cutting tools to carve his chair out of stone. I'm sure you get the idea.

Well, I think there is an application of another, similar as yet unnamed theory that could apply to certain popular concepts that catch the public's notice and then take on a life of their own. *Global warming,* is one of the newer ones while *civil rights* has been with us for quite a while. There are many others with more or less pizzaz. *The glass ceiling, Separation of church and state - weapons of mass destruction - the war on terror (drugs, poverty, ignorance etc.)* are others. Most will sooner or later take on political relevance which quickly distorts their reality and adds strong emotional fervor which can destroy any rational effort to study the problem or situation objectively and come up with realistic answers. Sometimes this is good–sometimes bad–either depending not on the realities of the situation, but on the political posture of those for and against any proposed solution. Often the factual realities of the situation become completely obscured by all the rhetoric which seems must be divided into for and against—a black or white rendering of virtually every expressed opinion most of which are actually infinitely varying shades of grey. It seems we always dichotomize—take one or the other of two opposing positions—either you are against it or for it. This emotional response completely denies reality.

For some time, the number one on the media dichotomy hit parade has been anthropogenic global warming (AGW). This includes the opinion that human contribution of carbon dioxide to the atmosphere is it's cause and must be dealt with. In the minds and voices of the media, many politicians, and significant world celebrities, anthropogenic global warming (AGW) is accepted as a proven fact, a serious danger to humanity, and a serious menace we must deal with. This has been so firmly established in the public's collective mind by politicians and their media supporters worldwide that any kind of discussion of other possible causes of global warming, is summarily lumped into their crackpot category. Their unrelenting, ego-driven conviction is that they are the only ones who have any understanding of this and everyone else is stupid or ignorant. This happens no matter how well documented by in-depth scientific studies, or supported by good scientific evidence that opinion may be. As Adolf Hitler once said, "If you tell a lie often enough and with great conviction, most people will believe it."

It's rather akin to a very old Chinese concept about proof of theories. They believed and stated, "Look only at data that supports the theory and ignore all data that does not." This was actually a practice of early Chinese theorists. It may well have been one of the reasons why Chinese science fell by the wayside after a very significant start. The Chinese no longer think that way. I hope we don't start to.

I read Al Gore's book and found it filled with hype and emotional appeal, slick photos and charts that struck an emotional response, but much of the science is one sided and thinly supported. To me it seemed more a *Chicken Little* book, long on emotion-stirring depictions and short on objective data. Incidently, some of the *data* in the book, particularly that in graphs, is absolutely impossible and provably so. I can see how it would appeal to those who want a simple direct answer and someone or something to blame for a causative factor that might or might not be relevant. James Hansen's article does some of those same things. Incidently, I do not question their intent to help solve a perceived problem. I do question the technical accuracy of their statements. Are they provable scientific data, or are they skewed to persuade?

Why We Can't Wait

by James Hansen
(I've added my comments in bold italics)1

There's a huge gap between what is understood about global warming by the relevant scientific community and what is known about global warming by those who need to know: the public and policymakers. (*Amen!*) We've had, in the past thirty years, one degree Fahrenheit of global warming. *(We actually had the same amount of cooling in a single year, 2007-2008.)* But there's another one degree Fahrenheit in the pipeline due to gases that are already in the atmosphere. And there's another one degree Fahrenheit in the pipeline because of the energy infrastructure now in place--for example, power plants and vehicles that we're not going to take off the road even if we decide that we're going to address this problem. *(There is little or no basis in fact for the last two statements. His one degree is a very arguable figure that has been estimated at between -1.8 and +1.9 by others in various studies by scientists of varying degrees of expertise and with varying agendas. So called "greenhouse" gasses are but one of several dozen significant factors affecting global climate. Most climatologists place a large number of these ahead of atmospheric carbon dioxide as having the most significant effect. For example, water vapor in the atmosphere has between 10,000 and 40,000 times the heat capacity of carbon dioxide at double present levels. Variations in water vapor literally overwhelms any greenhouse effect of carbon dioxide. Day by day changes in atmospheric water vapor are thousands of times greater in greenhouse effect than any change in carbon dioxide we could conceive of. Still, there is no doubt that man's contribution has at least some effect. Reduction in forests alone is one that could be bigger than CO_2. The question is, how much and is it significant enough to cause alarm. Pouring a cup of water into a fifty-acre pond may raise the surface level, but significantly?)*

The Energy Department says that we're going to continue to put more and more CO_2 into the atmosphere each year—not just additional CO_2 but more than we put in the year before. If we do follow that path, even for another ten years, it guarantees that we will have dramatic climate changes that produce what I would call a different planet—one without sea ice in the Arctic; with worldwide, repeated coastal tragedies associated with storms and a continuously rising sea level; and with regional disruptions due to freshwater shortages and shifting climatic zones. *(These are wildly speculative warnings at best. When he says, "we" is he speaking only of America, or the whole world? If this were a problem, it would be a global problem and if human generated carbon dioxide is a major cause, we have been moved into second place. China is now the one nation contributing the most carbon dioxide to the atmosphere. India, now in third place, is rapidly overtaking us. This is happening as we reduce our output while they are rapidly increasing theirs. So any real answer must be a global one that will work for and be shared by all nations. One big factor never mentioned is the effect of wholesale destruction of forests on the carbon dioxide balance. The burning to clear the land generates large amounts of CO_2, more than all the world's vehicles combined. Also, the removal of trees takes away one of the most efficient and effective systems to remove CO_2 from the atmosphere. Why don't we address that problem?)*

I've arrived at five recommendations for what should be done to address the problem. If Congress were to follow these recommendations, we could solve the problem. Interestingly, this is not a gloom-and-doom story. In fact, the things we need to do have many other benefits in terms of our economy, our national security, our energy independence and preserving the environment--preserving creation.

My book, "Energy, Convenient Solutions" describes many such scenarios. Some very real and practical ones that can be put in place relatively quickly and with all the possible benefits he describes. These practical solutions address numerous real problems that pose far more dangers than atmospheric CO_2.

First, there should be a moratorium on building any more coal-fired power plants until we have the technology to capture and sequester the CO_2. That technology is probably five or ten years away. *(I can't see anyone [and especially the Chinese] stopping the construction of coal fired power plants unless we rapidly go nuclear, geothermal, and wave action. Realistically, we will have to go to one or more of these sooner or later anyway. We have spent years searching for a practical method to sequester carbon dioxide emissions. We don't seem to be any closer to solving this knotty problem now than we were twenty years ago. I have been communicating with the MIT Laboratory for Energy and the Environment who have been trying to find a way to sequester carbon dioxide for many years albeit with very limited success.)* It will become clear over the next ten years that coal-fired power plants that do not capture and sequester CO_2 are going to have to be bulldozed. *(Will China do the same?)* That's the only way we can keep CO_2 from getting well into the dangerous level, *(We really have only wild guesses as to what constitutes a dangerous level, but taking that as a given, it certainly wouldn't do any harm to prevent further increases.)* because our consumption of oil and gas alone will take us close to the dangerous level. And oil and gas are such convenient fuels that they surely will be used. *(Who's he referring to? Most users of fossil fuels are located in countries where we can't tell people not to mine or use them)* So why build old-technology power plants if you're not going to be able to operate them over their lifetime, which is fifty or seventy-five years? It doesn't make sense. Besides, there's so much potential in efficiency, we don't need new power plants if we take advantage of that. *(Amen to potential! However, even with ultimate efficiency, carbon dioxide would still increase. What we really need is lots of wind, solar, geothermal, wave action, or safe, fourth-generation nuclear power. They are the only viable answers. France is building many new light water reactor power plants right now. Will we or China have the foresight to do the same? Or even better, develop viable wind, solar, geothermal and/or wave action power?)*

Second, and this is the hard recommendation that no politician seems willing to stand up and say is

necessary: The only way we are going to prevent having an amount of CO_2 that is far beyond the dangerous level is by putting a price on emissions. *(Not so! Far better would be to change over to the practical type of geothermal, wave action, or nuclear power generation and/or to a renewable fuel program I have described. As a result, use of fossil fuels and the associated emissions would disappear. Besides, the price he mentions, in the form of taxes, would mostly find their way into the political pipeline so politicians would be all for them.)* In order to avoid economic problems, it had better be a gradually rising price so that the consumer has the option of seeking energy sources that reduce his requirement for how much fuel he needs. And that means we should be investing in energy efficiency and renewable energy technologies at the same time. The result would be high-tech, high-paid jobs. And it would be very good for our energy independence, our national security and our balance of payments.

But a price on carbon emissions is not enough, which brings us to the third recommendation: We need energy-efficiency standards. That's been proven time and again. The biggest use of energy is in buildings, and the engineers and architects have said that they can readily reduce the energy requirement of new buildings by 50 percent. That goal has been endorsed by the US Conference of Mayors, but you can't do it on a city-by-city basis. You need national standards. The same goes for vehicle efficiency. We haven't had an improvement in vehicle efficiency in twenty-five or thirty years. And our national government is standing in court alongside the automobile manufacturers resisting what the National Research Council has said is readily achievable--a 30 percent improvement in vehicle efficiency, which California and other states want to adopt. *(Increased efficiency will not work in the long run because it only reduces emissions and could prove very costly. If atmospheric CO_2 is a problem, which I seriously doubt, stopping net emissions completely is the only* *viable, long range solution and the systems I propose do that.)*

The fourth recommendation--and this is probably the easiest one--involves the question of ice-sheet stability. The old assumption that it takes thousands or tens of thousands of years for ice sheets to change is clearly wrong. The concern is that it's a very nonlinear process that could accelerate. The west Antarctic ice sheet in particular is very vulnerable. If it collapses, that could yield a sea-level rise of sixteen to nineteen feet, possibly on a time scale as short as a century or two. *(This is no recommendation at all, but a description of a possible physical condition and its results. The complete melting of all the ice sheets in the world, arctic and antarctic, that are floating on the ocean will not cause any rise in the ocean level. That is a simple law of high school level physics. Only meltwater from ice in glaciers or that is otherwise supported by land masses will raise sea levels, ice such as that over central Greenland or Antarctica. I am much more concerned about the proven slowing of the gulf stream which is now flowing at less than half of what it did thirty years ago. Should it stop, as scientists believe it did during the so-called, "Little Ice Age," Europe would experience a drop in temperatures that would bring northern Scandinavian weather to England and much of Europe. At the same time, Greenland would experience a substantial increase in its ice pack. Then there are those reports of sea level rises from various places around the globe, mostly from islands. Many islands are actually sinking into the sea floor. Even parts of land on continents are sinking in some places and rising in others. This is caused by plate tectonic and other land movements. The resulting changes are responsible for the sometimes large increases and decreases in sea level reported)*

The information on ice-sheet stability is so recent that the Intergovernmental Panel on Climate Change report does not adequately address it. The IPCC process is necessarily long and drawn out. But this problem with the stability of ice sheets is so critical that it really should be looked at by a panel of our best scientists. Congress should ask the National Academy of Sciences to do a study on this and report its conclusions in very plain language. The National Academy of Sciences was established by Abraham Lincoln for this sort of purpose, and there's no reason we shouldn't use it that way. *(There have been a great many studies of ice, ice flows, glaciers, ice shelves [floating ice fields] and the ice cap on Greenland and Antarctica. Many of these are described in detail in Nigel Calder's book, "Miraculous Universe." Why don't we ask China to conduct such studies? They are now the largest contributor to atmospheric carbon dioxide and, unlike the US, they are rapidly increasing their emissions.)*

The final recommendation concerns how we have gotten into this situation in which there is a gap between what the relevant scientific community understands and what the public and policymakers know. A fundamental premise of democracy is that the public is informed and that they're honestly informed. There are at least two major ways in which this is not happening. One of them is that the public affairs offices of the science agencies are staffed at the headquarters level by political appointees. While the public affairs workers at the centers are professionals who feel that their job is to translate the science into words the public can understand, unfortunately this doesn't seem to be the case for the political appointees at the highest levels. Another matter is Congressional testimony. I don't think the framers of the Constitution expected that when a government employee—a technical government employee—reports to Congress, his testimony would have to be approved and edited by the White House first. But that is the way it works now. And frankly, I'm afraid it works that way whether it's a Democratic administration or a Republican one. *Politicians and political appointees are much more likely to pursue a course that promotes their political careers than one that properly informs the public. Even the dedicated scientists in federal agencies are certainly more likely to promote results that secure or advance their careers than that inform the public of unpopular results. So is the naked truth ever divulged by these individuals? Show me a person with no position to hold on to or no agenda to adhere to and maybe I would agree that, barring conceptual errors, the real truth could be forthcoming. The truth is, every individual has his or her own personal agenda guiding even scientific findings. Neglecting to mention a negative or a positive factor is a proven technique used to skew results to a desired conclusion.*

These problems are worse now than I've seen in my thirty years in government. But they're not new. I don't know anything in our Constitution that says that the executive branch should filter scientific information going to Congressional committees. *(Certainly this politically motivated implication intended to slam the executive branch is not born out by the facts.)* Reform of communication practices is needed if our government is to function the way our Founders intended it to work. *(How very true and nowhere is the need for such reform so evident as in the pronouncements of liberal Democrats and their supporters.)*

The global warming problem has brought an overall problem into focus. That is the pervasive influence of special interests on the functioning of our government and on communications with the public. It seems to me that it will be difficult to solve the global warming problem until we have effective campaign finance reform, so that special interests no longer have such a big influence on policymakers. *(One group's "Special interests" are another group's "concerned citizens." They each mean different things to different people and as far as I can see, virtually all politicians are controlled to some extent by some "special interests" called "concerned citizens" by those who benefit from their efforts. That our federal employees at every level, appointed and elected, are beholden to this group or that is obvious to all but the*

dimmest lightbulb among us. "Follow the money" may be a good axiom in searching for mischief, but "follow the chain of power and influence" should certainly be another.)

HoJo's further comments

My own opinion: Unlike statements I hear repeated like a mantra by so many of those riding the global warming wave, (a movement that has taken on the virtual trappings of a religion) this is not the warmest period on the planet in historical times. For obvious reasons I hold the motives suspect of those who repeat that mantra without ever even acknowledging all of the other plausible causes of climate change, or that it is not now as warm as it was a thousand years ago. Then there is the complete lack of acknowledging the fact that the atmosphere has been in a cooling trend since 1996.

In fact, the winter of 2012-13 has been the coldest ever recorded in most of the northern hemisphere. In addition, many locations report the heaviest snowfall on record. In Norway, flowering shrubs and fruit trees blossomed three weeks later than usual, the latest on record. I speak from experience as I was there at the time and saw what was happening and when. Will we hear anything about this from the *objective, unbiased* main stream media? Of course not. It would not sit well with their propaganda objectives.

It is a well-established fact that there was a period between about 600 CE and 1100 CE that was much warmer than the present. During that time crops and plants were grown in Scandinavia that now only grow in central Europe. Also during the same period, the Norse colonized Greenland, grew crops, raised cattle and sheep and built homes and churches. Then very suddenly, they, their crops and their animals disappeared completely, starved to death because the weather suddenly turned much colder. These are historical facts proven and confirmed by studies of ice cores and sea floor sediment. Greenland today is still unable to support farming and few of the plants that grew in Scandinavia then are present now. This is all born out by studies of pollen grains in bogs where levels relate directly to the times when the pollen grains drifted through the air. This is quite an accurate and proven technique used for many years to date levels of buried materials.

Another important factor that is noticeably absent from Al Gore's book, is that our measuring techniques have improved exponentially in both accuracy and detail in the last few decades. All measurements of temperatures and temperature variations before about 1960 are based almost exclusively on indirect measurements and anecdotal evidence. They are simplistic and questionable compared to the accuracy and in depth information now obtained by satellite and other technologies available in the digital age. Climate, global warming and global cooling (actually, a new ice age is now overdue) are far too complex with far too many variables to yield to our most sophisticated computers. When we are able to predict precisely where a hurricane is going to go and how strong it will become when it is merely a disturbance off the coast of Africa, then we may be able to say whether and how much global warming caused by atmospheric carbon dioxide is a reality with some degree of certainty. Now, no matter how you look at it, to say that the increase of atmospheric carbon dioxide is the responsible agent causing global warming in the manner it is being described by so many agenda driven spokespersons, is Chicken Little at best.

All that being said, the growing shortage of petroleum-based fuels coupled with their expanding use in China, India and other countries with rapidly expanding economies demands another easily transportable fuel system including portable power for trucks, busses, aircraft, and personal transport. That a by-product of this system is the reduction or complete removal of carbon dioxide emissions is a bonus that should make virtually everyone happy.

FLASH! Something new beneath the public RADAR: There is the uncomfortable fact that the US is now in the full bloom of a petroleum boom on top of a natural gas boom that is already well established. Both of these booms are being fed by new and improving mining technology primarily the fracking of shale deposits. The powerful potential of these booms is that the US could become energy independent in a few years, and even become an energy exporter. This could turn the US around economically to the dismay of leftists everywhere, even our own home-grown variety.

Links to several related web sites

There is considerable study about climate change conducted at MIT. I suggest a look at their web sites for some valuable information.

http://sequestration.mit.edu/research/survey2006.html to read the latest about the little known or reported techniques for sequestering carbon dioxide so it doesn't get into the atmosphere. Check the bottom of this site for links to more good information on the subject.

http://web.mit.edu/newsoffice/2006/survey.html to view the latest report on public attitudes about the environment and environmental action.

Both of these articles will be included in my book, *Energy and the Environment* if I get permission to quote them. Incidently, my book title may be changed to *SOLUTIONS*! I would quote them directly in my blog, but have not yet received permission to do so.

❖   ❖   ❖

Global Warming Report Gets U.S. Emphasis

By JOHN J. FIALKA - Thursday, Feb 15, 2007

WASHINGTON -- U.S. government scientists Friday said the long-term outlook for global warming may be more dire than suggested by this week's

United Nations' report, which they say doesn't fully address the impact of clouds and melting glaciers.

NOTE: for some very new and important information about clouds and their effects on climate, see the section on Henrik Svensmark's latest research starting on page 16 of this book.

Recent evidence of accelerated melting of glaciers in Greenland and the Antarctic ice cap came too late to be included in the report released Thursday by the U.N. sponsored Intergovernmental Panel on Climate Change.

Glaciers are among the largest sources of fresh water in the world and are contributing to rising ocean levels. Rising sea levels could expose population centers bordering the ocean to more storm damage and could require evacuation in some areas. But the computer models used for the IPCC report based their predictions only on the results of heating of the existing water in the world's oceans, causing the oceans to expand and sea levels to rise, said Tom Delworth, a climate modeler for the National Oceanic and Atmospheric Administration, the government agency in charge of climate science and weather service.

NOTE: Computer modeling has produced varying and unreliable results from lack of reliable programming, not to mention the hanky-panky discovered by the famous hackers. Creation of these modeling programs by those with an agenda provides highly skewed results that are frequently far off the mark and aimed to promote the programmer's position and opinions, not convey the truth.

The IPCC report predicts sea levels will rise by between one to two feet over the next 100 years. Mr. Delworth said there remains "much more uncertainty" over how much accelerated melting of glaciers might add to that.

A second area of continuing uncertainty has to do with the impact of clouds on climate change. Warming the ocean sends more water vapor into the air, and the resulting clouds accelerate global warming by trapping more of the sun's heat in the atmosphere and further warm the ocean. Jim Butler, deputy director of NOAA's global monitoring division, called this "a very scary feedback mechanism."

But, so far, the supercomputers the agency uses to model the effect on the earth's climate -- which were also used for the IPCC report -- aren't detailed or fast enough to predict how much clouds are accelerating the problem. Mr. Delworth said computer models divide the earth's oceans and atmosphere into four million boxes, each about 150 square miles, and that these boxes are too large to model the effects of clouds.

NOTE: the above paragraphs are nonsense. It is a well-established fact that clouds reflect infra red, the sun's heat, away from the planet and have a cooling effect on the atmosphere. The comments about clouds trapping heat are in direct contrast to the proven reality of increased clouds reflecting sunlight and heat back into space and reducing global air temperatures. In fact, increased atmospheric moisture caused by increased heat absorption by the atmosphere generates heavy cloud cover. This in turn reflects infra red from sunlight and acts as an effective control on global temperatures. This well known phenomenon prevents a runaway heating cycle which would long ago have made the earth too hot for life. This and Svensmark's Cosmic ray theory neatly explains the climate system changes for the last 500 million years. See the details in the section starting on page 16 of this book.

"We could use computers that are one million times faster than they are today and still not be satisfied," Mr. Delworth said.

How true! How true! I would change the word "satisfied" to "accurate," a very different meaning." Accuracy is what we need, not the satisfaction of those running the simulations. A revealing slip of the tongue by Mr. Delworth perhaps?

Further complicating the issue are layers of haze containing pollutants from human activity. Such pollutants, including sulfates, soot, dust and nitrates, tend to make the atmosphere brighter, reflecting more of the sun's heat back into space. *(Just like clouds do.)* The IPCC has found that the net effect of the added pollution is to cool the atmosphere.

A.R. Ravishankara, an atmospheric chemist for NOAA, said this raises a problem for governments attempting to clean the air by removing pollutants. "If you take away this cooling effect, then the heating effect would be exacerbated. It's a highly complex problem." Write to John J. Fialka at john.fialka@wsj.com

❖   ❖   ❖

Ancient Climate Change Flipped Ocean Circulation, May Repeat - Jan 5, 2006

SAN DIEGO, California (ENS) --> For the first time, evidence that climate change triggered a reversal in the circulation of deep ocean patterns around the world has been uncovered by scientists affiliated with the Scripps Institution of Oceanography. While the changes they describe occurred 55 million years ago, the scientists say today's conditions are similar and could have similar drastic effects on ocean circulation. In today's issue of the journal "Nature," scientists Fl via Nunes and Richard Norris describe how they examined a four to seven degree warming period that occurred some 55 million years ago during the closing stages of the Paleocene and the beginning of the Eocene eras.

"The Earth is a system that can change very rapidly," said Nunes. "Fifty-five million years ago, when the Earth was in a period of global warmth, ocean currents rapidly changed direction and this change did not reverse to original conditions for about 20,000 years."

The global warming of 55 million years ago, known as the Paleocene/Eocene Thermal Maximum (PETM), emerged in less than 5,000 years, an instant of geological time. Modern carbon dioxide input to the Earth's atmosphere from fossil fuel sources is approaching the same levels estimated for the PETM period, say the scientists, which raises concerns about future climate and changes in ocean circulation.

They say the Paleocene/Eocene example suggests that changes produced by human activities may have lasting effects not only on global climate, but on deep ocean circulation. Fossil records show that the global warming at the time of the PETM created changes ranging from a mass extinction of deep sea bottom dwelling marine life to migrations of terrestrial mammal species, as warm conditions may have opened travel routes frozen over when climates were colder. This period is the time when scientists find the earliest evidence of horses and primates in North America and Europe.

Nunes and Norris base their findings on the chemical makeup of microscopic sea creatures that lived 55 million years ago. The scientists analyzed carbon isotopes, or chemical signatures, from the shells of the one-celled animals called foraminifera, or "forams" that exist in vast numbers in a variety of marine environments.

"A tiny shell from a sea creature living millions of years ago can tell us so much about past ocean conditions," said Nunes. "We know approximately what the temperature was at the bottom of the ocean. We also have a measure of the nutrient content of the water the creature lived in. And, when we have information from several locations, we can infer the direction of ocean currents."

In the study, the scientists looked at a foram named Nuttalides truempyi from 14 sites around the world in deep-sea sediment cores retrieved via the Integrated Ocean Drilling Program, for which Joint Oceanographic Institutions, Inc., manages the U.S. component.

Chemicals from the foram's shells were used as nutrient "tracers" to reconstruct changes in deep ocean circulation through the ancient time period. Nutrient levels tell the researchers how long a sample has been near or isolated from the sea surface, giving them a way to track the age and path of deep sea water. Nunes and Norris found that deep ocean circulation in the Southern Hemisphere abruptly stopped the conveyor belt-like process known as "overturning," in which cold and salty water in the depths exchanges with warm water on the surface. Even as it was shutting down in the south, overturning appears to have became active in the Northern Hemisphere. The researchers believe this shift drove unusually warm water into the deep sea, likely releasing stores of methane gas that led to further global warming and a massive die-off of deep sea marine life.

"Overturning is very sensitive to surface ocean temperatures and surface ocean salinity," said Norris, a professor of paleobiology in the Geosciences Research Division at Scripps. "The case described in this paper may be one of our best examples of climate change triggered by the massive release of *greenhouse* gases and therefore it gives us a perspective on what the long term impact is likely to be of today's *greenhouse* warming that humans are causing."

Overturning is a fundamental component of the global climate conditions we know today, said Bil Haq, program director in the National Science Foundation's division of ocean sciences, which funded the research. Haq says overturning in the modern North Atlantic Ocean is a primary means of drawing heat into the far north Atlantic and keeping temperatures in Europe relatively warmer than conditions in Canada. Today, deep water generation does not occur in the Pacific Ocean because of the large amount of freshwater input from the polar regions, which prevents North Pacific waters from becoming dense enough to sink to more than intermediate depths. But in the Paleocene/Eocene deep water, formation was possible in the Pacific because of climate change, the researchers say, adding that the Atlantic Ocean also could have been a significant generator of deep waters during this period.

Warming climate slows ocean circulation 2-4-2006

Later this century, rising concentrations of *greenhouse* gases in Earth's atmosphere will slow the ocean currents that bring warm waters to the North Atlantic, thereby affecting that region's climate, computer simulations suggest. When the waters of the Gulf Stream and other warm currents of the North Atlantic reach an area south of Greenland, they cool, become denser, and sink. That, in turn, pulls more surface water northward, says Thomas L. Delworth, a climate scientist at Princeton University. The rate of this so-called thermohaline circulation depends on the temperature and salinity of the surface waters. The warmer and fresher those North Atlantic surface waters are, compared with underlying layers, the more buoyant they are and the slower the circulation becomes.

Using a new computer model, Delworth and his colleague Keith W. Dixon simulated various scenarios for ocean circulation in the North Atlantic from now until 2100. They calibrated the model using weather and ocean-circulation data gathered since 1860. Throughout the 20th century, rising concentrations of *greenhouse* gases such as carbon dioxide warmed the atmosphere and ocean surface, but not enough to slow the thermohaline circulation. That's because large amounts of air pollutants known as aerosols have scattered sunlight back into space and counteracted the *greenhouse* effect somewhat, says Delworth. In the remaining years of the 21st century, however, growing concentrations of *greenhouse* gases will begin to overwhelm the cooling effect of aerosols, Delworth and Dixon suggest. By the year 2040, thermohaline circulation could carry only 80 percent as much warm water to the North Atlantic as it does now. The researchers report their findings in the Jan. 28 Geophysical Research Letters.--S.P.

❖   ❖   ❖

SEA CHANGE IN THE ATLANTIC OCEAN? - April 26, 2004 By John Carey

One worry about global warming is that the increased concentration of *greenhouse* gases will upset earth's balance and bring changes in ocean circulation. In an extreme case, scientists say, the flow of warm currents up the Atlantic Ocean to Europe might be shut down. That would cause temperatures to plunge in Western Europe. Such a shift may be unlikely, but anxious researchers have been keeping a keen watch for any variations in ocean flow, using satellites and instruments moored out in the sea. Now, they are starting to spot some potentially worrisome changes. In the April 15 online issue of Science, a NASA-University of Washington team reports that the counterclockwise circulation of surface water in the North Atlantic has become markedly weaker since the early 1990s. "These observations of rapid climate changes during one decade may merit some concern," the authors write. But they also caution that it's not yet clear if the shift in circulation is the result of man-made global warming or part of a natural cycle.

THE GULF STREAM:

The North Atlantic: hot spot for ice ages

Warm waters linked to glacial eras - By Anthony R. Wood Jan 19, 2006

In this understated harbor village of tight streets and Cape Cod houses, the North Atlantic stirs gentle breezes in summer and tempers New England's harsh winter cold. And yet, only a geologic blip ago, this was a frigid and forbidding place, encased in a mile-thick sheet of ice. Massive ice sheets have advanced and retreated repeatedly over aeons, at a glacial pace. But what researchers have discovered recently is that climate can change in a hurry. Their findings have led to an ultimate irony: In the debate over global warming, one of the hottest issues is ice.

The planet's temperature has warmed robustly in the last 20 years. The year 2005 was the second-warmest year on record, and the Arctic polar cap is disappearing. *This statement is nonsense. There is no such cap. The Arctic ocean freezes to varying depths each winter and mostly melts away each summer. In any event the melting of any floating ice like that on the Arctic ocean does not raise ocean levels at all. That's one of the laws of physics.* The same melting that has raised concerns about rising sea levels has prompted counterintuitive scenarios that it could produce a fresh and disastrous big chill. Few foresee an imminent glacial outbreak, and some serious scientists insist that one is all but impossible, but ice-core records show clear evidence that rapid coolings and warmings have happened. And that was long before humans started burning the fossil fuels blamed for at least some of the modern warming.

Today, while the debate rages over how much humans are to blame for the planet's indisputable warming, scientists are still trying to figure out what conspired to bring on the flash-frozen ice ages. But a long and tortuous trail of evidence leads to a surprising suspect at the heart of the conspiracy: the Gulf Stream. Logically, it would be an unlikely culprit. It is hundreds of miles from the southern extent of the last ice sheet, and it covers only about 0.2 percent of the world's ocean surface. Yet the mighty stream is a critical piece of something much larger: the North Atlantic current system that moves warmth out of the tropics toward the North Pole and sends cold water back toward the equator, the so-called conveyor belt. It is estimated that the Gulf Stream transports about 20 percent of the heat moved by the oceans. If the Gulf Stream were to slow down or take a more southerly route, the change would disrupt the whole system, the North Atlantic would cool off. Europe and eastern North America might turn colder as the rest of the world heated up. Scientists think that's what happened the last time ice invaded Europe and the United States.

The question is: Could it happen again? In 1992, Richard Alley was in central Greenland, examining ice cores, when he saw something he could not believe. He and his colleagues were looking at the layers that told them about Greenland's temperature year by year, going back millennia. But instead of a gradual change, they saw radical shifts in the layers representing the climate 12,000 years ago. The temperatures had plunged and risen suddenly. He saw a swing of 15 degrees in a matter of 10, or no more than 30, years.

"This was a flipped switch, not a slowly turned dial," he recalls. "Something really dramatic had happened."

This made the *Little Ice Age* look like a snow flurry. Alley had come across a phenomenon described in 1985 by Wallace Broecker, a chemical oceanographer and paleontologist with Columbia University's Lamont-Doherty Observatory. Broecker called it the Younger Dryas period, for an Arctic shrub that mysteriously appeared throughout Europe. But whereas Broecker drew upon a variety of research sources, Alley was looking at direct physical evidence. The science of climate change was itself changing. Until the 1950s, climate was viewed as essentially a stable system, said Spencer Weart, head of the history center for the American Institute of Physics. That view was stood on its head when researchers saw evidence that big swings could occur in just a couple of millennia. By 1980, scientists came across further clues that such changes could happen in a few centuries. Broecker tightened the possible time frame in 1985 by publishing a paper on the Younger Dryas era. In the process, he indicted the North Atlantic and gave global warming an icon. The article, which appeared in the journal Natural History, posited that tundra conditions overspread Europe as the Gulf Stream and the North Atlantic heat-transport system broke down. Europe turned arctic.

That the North Atlantic would be so important underscores the complexity of oceanic circulation. The Pacific is triple its size, yet the Atlantic, Broecker explains, does a better job of moving heat northward than the wind-driven currents of the Pacific. And the mighty Gulf Stream is the engine driving it. A critical ingredient in the recipe for climate change is one of the most plentiful substances on the planet: salt. The key to keeping the conveyor belt in motion is the sinking action of the water. Salt adds weight to water, so the more saline it is, the better it sinks; the better it sinks, the faster the conveyor moves. Why is the Atlantic saltier than the Pacific? In part, said Broecker, it's because more fresh water from rain and snow drains into the Pacific than into the Atlantic. The differences are subtle but important. Every quart of ocean water has between 1.1 and 1.2 ounces of salt. Add a mere 0.03 ounces of salt to the water, and there is the same sinking effect as cooling the water by several degrees, by Broecker's calculation.

This is why any buildup of freshwater is so troubling: It could dilute the ocean subtly but critically. In the case of the Younger Dryas era, Broecker theorized that a mighty pulse of freshwater from melting glaciers stopped the sinking action. Freshwater accumulated in the far North Atlantic, and it froze. The conveyor suddenly slowed, interrupting the northward flow of warm water and warm air. The Gulf Stream couldn't do its job. What Alley found in his Greenland ice cores was that such a cosmic change could happen suddenly. In Weart's view, it marked a sea change in scientific opinion. "The whole notion of rapid climate change was very hard for science to accept," he said. "The guys who said there could be rapid climate change had to drag the rest of the climate community kicking and screaming." The chances of a shutdown of the conveyor are remote, if not out of the question. But any significant changes in the oceanic circulation would likely have major, and wholly unpredictable, effects on climate. At the peak of the *Little Ice Age,* the Gulf Stream did not shut itself down. Researchers think, however, that it slowed down, or maybe wandered from its usual trek.

If that happened again, they don't want to be caught by surprise.

Today, concerns about the state of the ocean run so deep that an unprecedented international effort is under way from the Straits of Florida to Greenland to track changes in the flow of the North Atlantic. So far, at least two new studies suggest that concerns about the freshwater buildup in the North Atlantic are warranted. Satellite data have detected a slowing of the circulation from Ireland to Labrador, according to a research team led by NASA's Sirpa Hakkinen. The team said that if the slowing continues, it might lead to large-scale ocean and, eventually, climate changes.

In a second study, 1,500 miles to the south, a group of British scientists reported in December 2005 a 30 percent slowdown in the movement of Atlantic deepwater. Hakkinen and Henry J. Bryden, the head of the British team, cautioned that their results weren't conclusive. Bryden looked at measurements taken at five intervals from 1957 to 2004. Hakkinen said it was impossible to predict whether the slowing in the so-called North Atlantic gyre would continue or was part of a natural cycle. Herein is a basic problem of oceanic research: the period of record is minuscule. Carl Wunsch, an oceanographer at the Massachusetts Institute of Technology, said researchers are beginning to build a baseline to track the movements of the North Atlantic conveyor. Right now, they have little basis for comparison. Climate evidently obeys the first rule of weather, only on a grander scale: What might happen is almost always more interesting than what is happening. If anything, however, the uncertainty makes it even more important to find out what the conveyor belt is up to. For the volatility of climate is inarguable. "These scenarios are conceivable," Wunsch said, "and we sure as hell want to know what's going on out there."

For more information on the gulf stream goto http//hjgulfstream.blogspot.com.

❖   ❖   ❖

An Inconvenient Truth or Convenient Distortion?

Thursday, July 27, 2006

My sister recently sent me a copy of Al Gore's book, "An Inconvenient Truth" along with the comment, "It is not a political statement, but a deep concern that I'm sure you have as well."

To so describe this book, which contains page after page of condemnation of the United States, capitalism, and the Bush administration, is at the very least, naive. I have read the book and yes, it does contain some true scientific basis for conclusions about the damage humanity is doing to our environment, but many of the conclusions are politically, rather than scientifically motivated. I have already expressed several opinions earlier in this blog which you can read merely by scrolling down through the various postings. Most important, I believe "An Unmentionable Menace!" to be a far more inclusive and impending danger than global warming can ever be. Global warming is but one small symptom of a whole series of things happening to our planet that portend great danger for humanity and indeed all life on the planet in the immediate future. Most of these dangers are completely ignored because they are far more "inconvenient" and politically unpopular, particularly with the left, than global warming. Mainly because politicians have yet to determined a way to use them to funnel money into their pockets.

Several interesting facts can be gleaned from Al's book. The graph on pages 33 and the explanation on page 32 clearly indicates that massive increases in vegetation (in the form of forests) could reverse the upward climb of atmospheric carbon dioxide. With a huge percentage of the earth's forests already turned into agricultural wasteland, primarily in the tropics, it is quite possible that deforestation and desertification have been the major factor in causing increases in atmospheric carbon dioxide. There is virtually no doubt about this. Unfortunately, those pushing for stricter governmental controls on emissions will completely ignore this major component cause of global warming since responsibility for this falls mostly on the third world. It is also interesting to note that there are only two major nations who have shown increases in forest since 1970. Those two are: Japan and the United States with the US showing the largest increase. This is not mentioned by those who seem to want to blame the US for all the world's problems including global warming. The truth of the matter is that the US has expended more practical effort to slow global warming and curb CO_2 emissions than any other large nation.

One glaring example of manipulation of data and grossly misleading information is shown on the graph of a thousand years of Northern Hemisphere Temperatures. All points before 1850 are based on completely different kinds of data than that after 1850 as evidenced by the sharp peaks and valleys after 1850. Also, since about 1930, there are data shown both above and below the zero line (both red and blue) at the same time. This is an absolute impossibility and calls the accuracy of the entire graph into question. Other graphs I have seen in Scientific American and evidence from the Viking's 500 year colonization of Greenland indicate that the climate in the northern Atlantic during the period from before 1000 CE until the mid-fourteen hundreds was probably even warmer than it is today. Northern Europe in particular was considerably warmer and supported more southern plants and animals during this period even than today. The Norse colonized two Greenland fjord systems in 984 and held this remote outpost of civilization for almost five hundred years before the onset of the "little ice-age" brought about their demise. They built a cathedral and churches, wrote in Latin and Old Norse, wielded iron tools, raised and stored large quantities of hay, herded farm animals, followed

European fashions in clothing–and finally starved to death and vanished.

The "Little Ice-age" brought about massive crop failures and much starvation in northern Europe as southern plants and animals retreated southward under the onslaught of continuing fiercely cold winters that continued until the mid 1800s. Scientists believe this phenomena was caused by the slowing or even the cessation of flow of the gulf stream in the Atlantic. It is interesting to note that scientists monitoring the Gulf Stream have noted it has slowed to about half it's normal rate in recent years and that, should it cease to flow, Europe could be in for another "deep freeze."

In the 1970s scientists expressed the same concern and issued warnings about global cooling and the onset of another ice-age. Information about this including dire warnings if we didn't do something about it were written in many of the same publications that now warn of global warming. The graph on pages 66 & 67 shows clearly the sudden reversal in CO_2 concentration at the end of each ice age coinciding with the melting of the glaciers and sudden rise in temperatures worldwide. Whatever part man's activities have in this scenario, use of fossil fuels may in fact pale in comparison to the destruction of forests that so efficiently remove CO_2 from the atmosphere. Pages 221 through 231 clearly indicate this problem. How this data can possibly be used to support use of fossil fuels as virtually the only cause of the rise in CO_2 and global warming is beyond my understanding. There is no question but that destruction of forests worldwide is a major factor in both CO_2 increases and global warming. Unfortunately, since that part of the problem falls clearly on third world nations and not the U.S. it is never mentioned.

Pages 240 through 245 likewise are not indications of the effects of atmospheric CO_2, but of our diversion of rivers which creates desert like conditions in many areas, adding to local warming. The direct actions of mankind, not global warming, is responsible for these tragedies. Likewise, the melting snows of Kilimanjaro were not caused by global warming as indicated on pages 42 through 45, but by the devastating cutting of forests around the lower parts of the mountain. This reduced the airborne moisture that once fed the snows on the mountain. In 2004 the British science journal Nature noted this as a fact. In 2004 this was confirmed by major studies reported in the International Journal of Climatology and the Journal of Geophysical Research. They showed that the loss of snow was not caused by global warming, but by the aforementioned deforestation. Mr. Gore's implications were precisely the opposite of the truth.

The Kyoto treaty adhered to by all nations would be an economic disaster for the U.S. and would do virtually nothing to even slow global warming. As environmentalist Peter Roderick states, "I think everybody agrees that Kyoto is really, really hopeless in terms of delivering what the planet needs." Add to this the statement of former U.S. Senator Tim Wirth of Colorado, "We've got to ride the global warming issue. Even if the theory of global warming is wrong, we will be doing the right thing–in terms of economic policy and environmental policy." That certainly describes the true agenda of the fundamentalist church of AGW so even the dimmest lightbulb can understand it. The policies he refers to are those of the radical left who are doing all they can to gain more and more control over the lives of the American people.

One thing Al got almost right is expressed on pages 214 through 220 – the population explosion. I think the grossly optimistic line of blue on the graph on page 217 shows an unwarranted turn to the right around 2050. I see no indication the population will level off at that number except for massive starvation, murder and mayhem all over the world. Sadly, this may turn out to be the case. A glance through the earlier parts of this blog deals with this in depth.

Again, this is not an effect, but certainly is a part of the cause of any anthropogenic global warming.

Yes this book and probably the movie contains facts which point to problems facing humanity. Unfortunately, distortions and political hype are what the left wing socialists and their cohorts in Hollywood and the media will concentrate on with scare tactics to gain political power and control, and to promote their own anticapitalist, anti-conservative, anti-America agendas.

The following quote sums it up quite fairly, "Improving the environment requires engineering, scientific and economic competence and involvement. The idea that a return to sustenance, communal living is a corrective measure is pure nonsense. Improvements in the natural environment, improved food production and improved quality of life for humans occur only where there is capitalism and the use of synthetic chemicals. In the USA forests have increased by 140 million acres since 1920 with accompanying increases in bird and animal life, and a decrease in soil erosion. (Not to mention the removal of huge amounts of carbon dioxide from the atmosphere.) Third World nations (primarily in Africa, Asia and South America) destroy forests only because they lack alternative fuel/energy sources." (And because of excess population, corrupt socialist dictatorships and self-serving leaders.)

❖   ❖   ❖

The Substantial Benefits of Trees

Trees provide a multitude of benefits. Unfortunately, much of the general public is not well informed on this topic. By increasing awareness of the benefits relating to trees, we can all utilize current scientific evidence to help resolve many challenging issues and improve the livability of our cities. Proper tree care and sound forest management programs are crucial to the health, longevity, and sustainability of

our urban forests. The care of trees is a wise investment in our future.

A list of the benefits of trees would include at least the following:

Air Temperature and Energy Consumption

Trees cool air temperature and help to offset the "heat island" effect of hardscapes (formerly forested areas now occupied by buildings and pavement) by providing shade and by transpiration (the release of water vapor into the air). By properly selecting and planting trees, yearly energy savings can exceed 40%. Three mature shade trees placed strategically around a house can cut air conditioning bills by 10% to 50%.

A single large tree can release up to 400 gallons of water into the atmosphere each day. Water from roots is drawn up to the leaves where it evaporates. The conversion from water to gas absorbs huge amounts of heat, cooling hot city air.

Dallas area neighborhoods with mature trees can be up to 11 degrees cooler than neighborhoods without trees. A one-degree rise in temperature equals a 2% increase in peak electricity consumption.

One simulation found that planting 500,000 trees in the Tucson area would lower the "heat island" effect by 3 degrees and overall cooling costs by up to 25%.

Cities are 5 to 9 degrees warmer than rural areas and 3% to 8% of summer electric use goes to compensate for this urban "heat island" effect.

The National Arbor Day Foundation calculates that 100 million additional mature trees in U.S. cities would reduce the "heat island" effect and save $2 billion annually.

Air Quality

Trees produce oxygen and store carbon dioxide (the opposite of humans), which helps to clean and restore our air. The American Forests organization's studies foresee the value of the urban forest to U.S. cities to be $10 billion by storing carbon dioxide, cleaning particulate matter, and generating oxygen for urban spaces.

One acre of trees produces enough oxygen for 18 people every day.

One acre of trees absorbs the carbon dioxide produced by driving an automobile 26,000 miles.

A fully-grown Sycamore tree can transform 26 pounds of carbon dioxide into life-giving oxygen every year.

Large trees remove 60 to 70 times more pollutants than small trees. Only a small portion of the Dallas area tree population exceeds 24 inches in diameter.

For every ton of wood an urban forest grows, it removes 1.47 tons of carbon dioxide and replaces it with 1.07 tons of oxygen.

A typical tree removes 25 to 45 pounds of carbon from the air each year.

A study of Atlanta's urban forest showed that intense urban development and subsequent removal of large urban forest areas increased the "heat island" effect. This raised the levels of isoprene emissions, increasing the natural formation of harmful ozone.

An EPA study in Chicago showed that the 23.2% of canopy cover in the Lincoln Park neighborhood adjacent to downtown annually filters 43.9 tons of particulate matter, 14 tons of carbon dioxide, and 12.4 tons of nitrogen oxides, giving the urban forest an estimated pollution abatement value of $625,000 per year.

Water/Soil

Planting trees along streams, wetlands, and lakes, helps control storm water runoff, removes soil sediment, reduces flood damage, and increases water quality, by reducing the pollution of the water runoff by as much as 80%.

Healthy, vegetated stream buffer zones reduce the total suspended solids phosphorus, nitrogen and heavy metal transfer between urban areas and streams by 55% to 99%.

Numerous studies show that trees along streams increase fish populations.

The urban forest reduces erosion. One square mile of forest land produces 50 tons of erosion sediment. In contrast, farmland produces 1,000 to 50,000 tons, and land prepared for construction produces 25,000 to 50,000 tons of sediment per year.

Tree canopy, in one study, reduced surface runoff from a one-inch rain over a 12-hour period by 17%.

In natural watersheds with trees and vegetation, 5% to 15% of stream flow is delivered as surface storm water runoff. In highly developed areas, more than 50% of stream flow is delivered as surface storm water runoff.

Animal Habitat

Trees attract wildlife to an area by supporting habitat and creating biodiversity.

Trees provide food and shelter for wildlife.

Economics, Health, and Psychological and Social Behavior

Trees offer unlimited climbing challenges and good physical activity opportunities such as tree swings and tree houses.

Numerous trees and plants have proven useful in phytoremediation or removal of toxic materials from soils.

Trees can become living witnesses to our history and evidence of our cultures. Without a cultural history, people are rootless. Preserving historical trees offers lingering evidence to remind people of what they once were, who they are, what they are, and where they are. Trees feed our sense of history and purpose.

Studies across the nation show that residential home prices increase from 3% to 20% due to the presence of trees, depending on the type of trees, scarcity of treed lots, and the maturity of existing trees.

One widely reported study showed that viewing trees through a window during surgery recovery cut the average recovery time by almost one whole day compared to patients with a view of a blank wall.

People turn to the urban forest, preserved by humans as parks, wilderness, or wildlife refuges, for something they cannot get in a built environment. The quality of human life depends on an ecologically sustainable and aesthetically pleasing physical environment. The surge of interest in conserving open spaces from people motivated by ecological and aesthetic concerns is growing.

Trees curtail health care costs by facilitating positive emotional, intellectual, and social experiences.

Environmental stress may involve psychological emotions such as frustration, anger, fear and coping responses; plus associated physiological responses that use energy and contribute to fatigue. Many who live or commute in urban or blighted areas experience environmental stress. Trees in urban setting have a restorative effect that releases the tensions of modern life. Evidence demonstrating the therapeutic value of natural settings has emerged in physiological and psychological studies. The cost of environmental stress in terms of work-days lost and medical care is likely to be substantially greater than the cost of providing and maintaining trees, parks, and urban forestry programs.

Trees are a source of food for humans, i.e. nuts, apples, stone fruits like peaches and apricots, citrus, avacados, etc. On a large scale, trees require less fertilizer and keep the soil healthier than most crops.

Aesthetics

Trees can screen objectionable views, offer privacy, reduce glare and light reflection, offer a sound barrier (acoustical control), and help guide wind direction and speed.

Trees offer aesthetic functions such as creating a background, framing a view, complementing architecture, and bringing natural elements into urban surroundings.

Here are links to sites with much accurate information.

Http://www.fao.org/DOCREP/005/AC805E/ac805e0s.htm

Trading forest carbon to promote the adoption of reduced impact logging

Joyotee Smith and Grahame Applegate*

* Center for International Forestry Research (CIFOR), Jl. CIFOR, Situ Gede, Sindang Barang, Bogor Barat 16680, Indonesia, Tel. +62 (251) 622 622. Fax +62 (251) 622 100, E- mail: e.smith@cgiar.org and href=”mailto:g.applegate@cgiar.org”>g.applegate@cgiar.org

http://cordis.europa.eu/euroabstracts/en/october01/feature01.htm

Net Biome production of managed forests in Japan.

http://www.the-tree.org.uk/TreeTalk/News/newsarchive.htm

FAQ Global changes

http://www.newton.dep.anl.gov/askasci/gen01/gen01491.htm

Ask a Scientist (Argonne National Lab)

❖   ❖   ❖

Volcano in Iceland spews millions of tons of CO₂ into the air

Okay, here's the bombshell. The volcanic eruption in Iceland, since its first spewing of volcanic ash has, in only **FOUR DAYS OF ERUPTING, NEGATED EVERY SINGLE EFFORT** you have made in the past five years to control CO₂ emissions on our planet, all of you.

AGW promoters read this . . . and weep!!!

Professor Ian Plimer (a member of the School of Earth and Environmental Sciences at the University of Adelaide. He is also a joint member of the School of Civil, Environmental and Mining Engineering) could not have said it better! If you've read his book you will agree, this is a good summary.

Of course you know about this evil carbon dioxide that we are trying to suppress, that vital chemical compound that every plant requires to live and grow, and to synthesize into oxygen and plant food for all animal life including humans. In fact the increase in CO₂ is already responsible for an increase in crop growth of from 15% to 40%. (See page 37)

I know, it's very disheartening to realize that all of the carbon emission savings you have accomplished while suffering the inconvenience and expense of: driving Prius hybrids, buying fabric grocery bags,

sitting up till midnight to finish your kid's "The Green Revolution" science project, throwing out all of your non-green cleaning supplies, using only two squares of toilet paper, putting a brick in your toilet tank reservoir, selling your SUV and speedboat, vacationing at home instead of abroad, nearly getting hit every day on your bicycle, replacing all of your 50 cents light bulbs with $10.00 light bulbs . . . well, all of those things you have done to lower your carbon footprint have gone down the tubes in four days.

Gasses and the volcanic ash emitted into the Earth's atmosphere in four days - yes - FOUR DAYS ONLY by that volcano in Iceland, has totally erased every single effort you have made to reduce the evil beast, carbon. And there are around 200 active volcanoes on the planet spewing out this crud at any one time - EVERY DAY.

I don't really want to rain on your parade too much, but I should mention that when the volcano Mt Pinatubo erupted in the Philippines in 1991, it spewed out more *greenhouse* gases into the atmosphere than the entire human race had emitted in its entire YEARS on earth. Yes folks, Mt Pinatubo was active for over one year, think about it.

Of course I shouldn't spoil this touchy-feely tree-hugging moment and mention the effect of solar and cosmic activity and the well-recognized 800-year global heating and cooling cycle, which keep happening, despite our completely insignificant efforts to affect climate change.

And I do wish I had a silver lining to this volcanic ash cloud but the fact of the matter is that the bush fire season across the western USA and Australia this year alone will negate your efforts to reduce carbon in our world for the next two to three years. And it happens every year.

Remember that your government tried to impose a whopping carbon tax on you on the basis of the bogus "human-caused" climate change scenario.

Hey, isn't it interesting how they rarely mention ''Global Warming'' any more, but now call it ''Climate Change'' - you know why? It's because the planet has COOLED by 0.7 degrees in the past century and these global warming bull artists got caught with their pants down.

And keep in mind that you might yet have an Emissions Trading Scheme (that whopping new tax) imposed on you, that will achieve absolutely nothing except make you poorer and make a few politicians a great deal richer. It won't stop any volcanoes from erupting, that's for sure.

But hey, relax, give the world a hug and have a nice day!

PS: I wonder if Iceland is buying carbon offsets?

NOTE: The amounts of CO_2 released by volcanos as described in this section by Plimer is disputed by numerous of his detractors. I don't know who is correct as I have no way of checking their calculations. My guess is the real numbers lie somewhere in between the two, as both have their own agendas to serve.

❖   ❖   ❖

How about Some Real Answers?

Some real, accurate observations about the bigger picture of environmental destruction of which anthropogenic climate change could be but a tiny, insignificant part - Friday, June 23, 2006

Nowhere in most articles about global warming, including the one I urge you to read following this section, is anything ever mentioned or suggested other than conservation of energy. There is rarely, if ever any real solution to the problem even hinted at. In the article by Paul Loeb, there was not a single reference to any serious effort at solving the problem - none!

Other than the obvious political thrust of the article, conservation is all that is ever mentioned.

China and India, two of the most populous nations in the world, are rapidly accelerating their energy use as their burgeoning economies expand rapidly. These two nations each promise to dump far more CO_2 into the atmosphere and dwarf any contribution from the US in a very short time. They are exempt from the Kyoto protocol. In both nations, pollution has already become overwhelming and deadly.

There are real, practical solutions. Not the government or the many university and public research groups, but the free-enterprise capitalist organizations in the US have it within their power to solve this problem and solve it quickly. No, not by stopgap measures such as conservation, or very long range, pie-in-the-sky proposals like the hydrogen fuel-cell vehicle, but with real solutions that provide nearly boundless energy, fixed and portable, for buildings and vehicles and with very little modification to power plants large and small. All this while putting a complete stop to increases of atmospheric CO_2. We have the technology to do it right now! We have the raw materials at hand. We have the people who know how to design and produce the infrastructure, the power plants, the vehicles and their engines large and small. They are already here and working! A complete changeover can be handled in as little as ten years.

The benefits to our nation and to the world are staggering: no drilling in pristine places, no imported petroleum, no huge investment in a radical new distribution system, no obsolescence of current vehicles, no drain on our resources as we send increasing billions to oil-rich nations most of who's leaders are working to destroy us. Instead we are looking at numerous and widely spread small fuel producers, producers of safe and efficient fuel out of totally renewable resources - right here in out own country. Many more high-paying jobs right here! The billions now going to the middle eastern nations will stay right here at home. The tools, challenges, answers, and obstacles to this beneficial effort are detailed in a book titled, *Energy, Convenient Solutions* by Howard Johnson. This book is currently available from most book sources and as an ebook for the Barnes & Noble Nook. It is also available from AKWBooks and for Amazon's Kindle. It is available directly from the publisher by email **senesisword@yahoo.com**. This amazing compendium of answers to our growing energy crisis was published in February 2011.

To read excerpts of the book goto:
http://senesisword.blogspot.com

For a review of the book, goto:
http://ecsreview.blogspot.com

❧   ❧   ❧

Another, far more important point:

I have for at least twenty years given lectures on a far more comprehensive problem of which global warming is but a very small part. The name of this lecture is, *"The Decimation of the Environment - The Real Culprit."* This is the real, root cause of virtually all of our growing environmental problems and is what will ultimately bring about our total demise and the collapse of the entire world of human societies.

Text of the current lecture can be read by selecting: http://decimatenviro.blogspot.com

My son gave me a book for Father's Day entitled, *Collapse - How Societies Choose to Fail or Succeed* by Jared Diamond, Pulitzer Prize winning author of *Guns, Germs, and Steel.* This book, published in 2005, says, in fine detail, how and why past societies have died out, literally. It also projects where we are headed. It details precisely what I have been saying in my lecture for at least twenty years even using one of the same examples I have used in these lectures, Easter Island.

To my mind, this book is far far more important and definitive than Al Gore's book or movie, yet I can guarantee it will not get but a tiny fraction of the attention that they get. This is probably because it is not political and isn't primarily a **bash someone for political purposes** book. At the back of the book are listings of many articles, studies and references to more in-depth information relating to the book's content.

If you want a real, highly definitive look at our earth, society, where we are headed environmentally and what we can realistically do and expect, I urge you to get and read this book. It doesn't blame George Bush or Bill Clinton or liberals or conservatives for mankind's insatiable appetite for sex and procreation that is inundating Europe with Muslim immigrants and the US with Latino immigrants. It does describe the human foolishness that has, in the past, brought about the cruel death of all individuals in certain societies and draws the conclusion that, unless we do something immediately, the whole of mankind could suffer the same cruel fate and in the **very near future**.

Unfortunately, not only will mankind be cruelly eliminated, but most other life on earth as well.

❧   ❧   ❧

Decimation of the Environment
The Real Cause of the Growing Environmental Disasters

Lecture given by Howard Johnson to a group of Scientists and Engineers in South Bend Indiana, Tuesday, September 13, 2005 - revised 2010

I am going to talk to you about a very controversial subject. I hope to open your eyes and minds to a very real and present menace facing us all. While we direct increasing efforts to deal with the numerous symptoms of this menace, almost none is directed at the root cause. I may seem a bit crazy to some of you, but to those I address the following words by Angela Monet:

"Those who dance are thought insane by those who can't hear the music."

During 1953, while living in Long Beach California, I took my Professional Engineering qualifications training and examination. As part of our training, my group of six hopeful Professional Engineers were responsible for developing projections of population, food supply, fuel reserves, atmospheric conditions, raw material usage, technological advances and a few other conditions for the next fifty years of our planet's life. I ran across my copy of the resultant predictions while moving my personal belongings in 1981. I was startled at what I reviewed in those papers. Some of our predictions were uncannily accurate while some were far off the mark. The projections of fuel reserves and raw material usage were very accurate. Our population growth projections were a bit under the mark and those for our atmospheric conditions were over for some components and under for others. We grossly underestimated the speed of developments in rocketry, communications and computer technology based on silicon chips. We did not foresee the rapid decimation of the rain forests and the virtual extinction of so many wild animals, particularly the predators. Our projections of food crops were under-estimated while the collapse of the ocean fisheries was completely missed.

I have continued to read widely regarding these subjects from research papers, books and magazines written by scientists from universities and creative thinkers from all over the world. Most of these works direct answers to the many vital and controversial questions concerning the actions of the human population and their effects on our earth. This accumulation of facts points unerringly toward some disturbing conclusions about the very near future of our planet. These conclusions are extremely controversial and frightening and most people refuse to even listen to them let alone do anything about them. As part of the process of writing a book on this general subject I am accumulating those facts and ideas which cause my concerns. A small portion of these facts, concerns and conclusions are included in this essay. The opinions are my own, the facts I report are as accurate as possible.

THE DISTURBING FACTS:

POPULATION GROWTH: The human population continues to grow explosively. Despite the expressed concerns of a tiny group of people, there is no indication of any change in this growth. Currently, we are adding about 200 million people each year to our already crowded planet. That is about twice the population of Mexico. The only country spending any effort at curbing population growth is Communist China and they have been denounced by many for their efforts. In many nations with high birthrates, population pressures cause their excess population to emigrate or die of starvation. As they move to other nations, they bring their high birthrates with them and quickly overwhelm the resources of these other nations. Europe and the Middle East as well as The United States and Latino nations like Mexico are classic examples. Most Moslem nations have a birthrate among the highest in the world at 4.6 per woman. Latino nations of the Americas have similar birthrates. As a result, poor families are emigrating to Europe and the US at astounding rates.

The Lemming Syndrome

Lemmings are small rodents, usually found in or near the Arctic, in tundra biomes. They are subniveal animals, and together with voles and muskrats, they make up the subfamily Arvicolinae (also known as Microtinae), which forms part of the largest mammal radiation by far, the superfamily Muroidea, which also includes rats, mice, hamsters, and gerbils. Lemmings are best known for their mass migrations caused by their rapid reproductive rate and overpopulation.

Simply stated, all lemming populations expand in their home area until their density overruns their food supply. At this critical point, they begin to migrate outwards from a single focus area. As the migrants enter the territory of other lemmings the critical point in their food supply and population density causes even greater and more rapid migration into new territory expanding the area of over critical density. Because lemmings are good swimmers, the waves of migrating lemmings moves quickly across ponds. streams and rivers. Should they enter water too wide for them to swim across, they eventually tire and drown. This is quite similar mathematically to a nuclear chain reaction in an atomic bomb. This happens in four year cycles in Norway, thus the myth that they commit suicide.

Once the waves of emigrating lemmings subside, the population density drops precipitously, and the food supply is able to replenish, the remaining lemming begin the cycle all over again. While predation has some effect and predators multiply along with the lemmings, they too are subject to fluctuations in population along with their food supply. It's similar to the population fluctuations of populations of the snow shoe hare and the lynx.

It doesn't take any stretch of the imagination to equate the lemming population changes to that of humans at the present. Consider the huge tide of humanity that has been crossing our southern border both legally and illegally for many years. The Latino population of the US is rapidly surpassing the African population in numbers. At the present rate of immigration and high birth rate, Latinos will be a true majority in the US by the year 2030 when the US population exceeds a half billion! To survive we will become a net importer of food rather than the huge exporter we are today. Where will that food come from?

Several European nations are now approaching Moslem majorities and those Moslems are becoming more and more militant as their numbers grow. With population growth rates several times those of native Europeans, Moslems are rapidly approaching the day when they will place their Mullahs in political control of those nations. England and France will probably be the first to become Islamic dictatorships where everyone is ruled by Sharia law and religious freedom disappears. Think Iran or Syria. Consider this the lemming syndrome applied to humans.

If you think humanity is decimating the environment now, consider what Moslem nations under the irrational behavior of Sharia law will do, especially when food comes into short supply.

FOOD PRODUCTION: World human food production peaked during the nineties and is now declining. There are many contributing factors including the following ones:

1 All ocean fisheries, except the Indian Ocean, are declining. In fact, the North Atlantic area fisheries have virtually collapsed and will take many decades, perhaps centuries, to restore, even under the best of controlled conditions. The use of new technologies has enabled commercial fishermen to decimate the most productive fish populations with uncontrolled harvesting. While there is a tentative world accord aimed at this problem, the worst offender nations have not signed the accord and continue to plunder the oceans with ever expanding fishing fleets harvesting far more than the fisheries ability to renew the resource. A byproduct of this gross harvesting of wild fish for food is the destruction of sea floor habitats by heavy bottom nets and the killing of millions of tons of non food sea creatures. Add to this the destruction of many areas where ocean fish spawn and their young can grow, protected from predation. Wild Atlantic salmon and cod fisheries in particular have declined almost to the point of being non existent.

2 Agricultural lands are now shrinking in area for the very first time. The clearing of forests while decimating wildlife habitats is now generating less agricultural land than that which goes out of production due to soil depletion, erosion, and desertification.

3 Uses of chemical fertilizers, insecticides, weed killers and fungicides have reached a point of diminishing returns while polluting our watersheds. Harmful pest populations continue to develop immunities to these poisons much faster than do beneficial populations of insects, birds, plants and other natural pest controls which are higher up the food chain.

4 Slash and burn agriculture, used throughout the third world, is rapidly destroying not only habitats but the usefulness of much land for any purpose whatever! The soil is usually depleted after a few years use to where it will not produce crops. It then takes decades or even centuries for nature to return the soil to productivity again. A byproduct of this denuding of the land is air pollution and massive erosion with the silting of rivers and mud-slides of monstrous proportions. TV news programs increasingly show this happening all over the world. Many of these reports show bare ground, air so polluted by smoke it is dangerously unhealthy with low visibility over many square miles, and rivers so muddy and/or polluted all life is choked from them. Acid rain is but one more negative factor in a sea of bad news.

5 Fresh water supplies are shrinking all over the planet. Underground water reservoirs are being depleted drastically as water is mined much faster than it can be replenished. Rivers, damned for irrigation of marginal land, are frequently drying up before the water reaches their termination in lake or sea. Everywhere, lakes and wetlands are shrinking or drying up completely. Mexico City was once called the Venice of the Americas for its many lakes and canals. The city has so drained its aquifer that the lakes and canals are long gone and the land itself has sunk more than twenty feet. The shifting ground has broken so many of its water mains that leaks now waste up to a third of its water. Subsiding ground is a serious problem in many other parts of the world as

well. Americans mining the water of the Ogallala reservoir underlying the great plains must go deeper and deeper as this huge reservoir shrinks. Already it is gone from parts of Texas. In China and India, some underground aquifers have shrunk so much water can no longer be reached and food production suffers. The Yellow River in China slows to a trickle before drying up because of upstream irrigation. This is also true of the once powerful Ganges, Nile and Colorado Rivers which barely reach the sea in dry seasons. Literally thousands of small rivers have completely disappeared. Tule Lake in California's San Joaquin Valley, once more than 100,000 acres is now less than one tenth that size and is little more than a knee-deep mud puddle. The diversion of water for agriculture has destroyed 90 percent of California's wetlands and caused the extinction of up to 39 of 67 native fish species. This trend is echoed world wide and is another way natural habitat is increasingly being destroyed.

These are all valid demonstrations of how finite resources are reacting to uncontrolled human population growth. Can anyone hear and recognize what is going on and not be terrified?

NOTE: there is one bright spot in the food situation and it bears directly on man's addition of CO_2 to the atmosphere. From a scientist who's specialty is the effects of increased CO_2 on crop plants.

He reports, "The proven beneficial effects of increased CO_2 means in turn that something like a billion humans are being fed by the EXTRA crop growth from CO_2. This amounts to an increase in plant growth of from 15% to 40% depending on the plant species. So the contributions of CO_2 to the atmosphere by man's activities is having a very positive effect on the human food supply.

"Just think of the harm -- the starvation -- if the extra CO_2 could be instantly made to go away, as the catastrophists apparently wish. It is not a pretty scenario at all."

See the entire story on page 37.

❖   ❖   ❖

THE EPIDEMIC THAT MOST MENACES THE EARTH

I received a copy of a commentary on sustainability on our planet by Jay Burney. His comments fit so well with my passion about population, I have included a few quotes in this piece. After the quotes, I will provide a number of responses. I do not disagree with Mr. Burney in the main. I strongly disagree with many of his comments about the causes and what needs to be done to change things. I see his efforts as reactions to symptoms and condemnations of those who see things differently. I see no one addressing the real problem with more than the most casual comments. Some of his comments follow:

January 2002 - In 1992 a report was issued by the Union of Concerned Scientists called *World Scientists Warning to Humanity*. It stated bluntly that *human beings and the natural world are on a collision course.* It says that our current practices and activities are altering the world in ways that make life unsustainable. The report was signed by over 1700 scientists representing 77 countries and by over half of the world's living Nobel laureates.

A few years before that, in 1987, The World Commission on Environment and Development chaired by Norwegian Prime Minster Gro Haarlem Brundlandt issued a report called *Our Common Future*. That report provided a definition of sustainability that stands today as a blueprint for thinking and action. It states that *Sustainable development is development that meets the needs of the present without compromising the ability of future generations to meet their own needs.*

That is a pretty good definition. It is not that complicated to understand. It is about learning to live within our means. And so we know that we have problems, and that we as humans, need to change how we live on this planet. Of course the devil, as they say, is in the details.

NOTE: Our President and the Congress should take note of the above comment, but probably will ignore it. They are driving a juggernaut of economic destruction being used to give them ultimate power while destroying freedom in America. Many voters are too ignorant, stupid, or self-serving to care.

In Burney's commentary he issued many statements including the following phrases:

globalization is defined by powerful **economic** interests

an approach to **business** and economic growth, that dictates that the bottom line is to make as much profit as possible at all costs.

Growth is promoted as good for **business**. And we are told, what is good for **business** is good for society.

This approach fails to consider transcendent environmental and social costs including clean and renewable resource use, healthcare, hunger, and education.

This myopic philosophy of **business** leads to environmental depredation, social unrest, and a widening of the gap between the rich and the poor.

While it may serve the short-term needs of a few, it fails to even suggest sustainability

Businesses that put profit over social responsibility, and that do not recognize the true costs of environmental degradation are looting the natural wealth of the planet.

Businesses that don't recognize this are stealing our future. We know that now it is the time to recognize that the real bottom line is the environment.

Unchecked human population pressures are contributing to environmental, social, and economic problems. The world population in 1960 was 3 billion. Today it has doubled to a little over 6.2 billion. It will double again in less than 50 years.

We don't know how to feed the worlds population today. What will it be like in another generation? Two generations. We can only imagine what it will be like in a hundred years.

We do know that globalization has increased economic activity worldwide, soaring last year (2004) to an estimated $30 trillion. We also know that it has increased income inequality and environmental degradation.

I believe his opinions and attacks on **business** reflect a type of bigotry that would not be tolerated if it were directed toward a racial or ethnic group. Business, like all other organized human efforts, including religions, political organizations, governments, social groups, occupations, professions and countless others, is a grouping of people with a common denominator. Many individuals belong to a number of these groups and can be identified as such. No one belongs to less than four including: sex, race or ethnicity, economic level, age and there may be more. As such, all groups of any kind have members ranging through many spectra: intelligent to stupid - belligerent to friendly - leaders to followers - educated to ignorant - power hungry to meek - rich to poor - humanitarian to sociopath, and countless others. This is true for groups as diverse as families, governments, businesses and churches.

I would edit this piece by placing politicians driven by personal gain ahead of businesses as they are certainly more self serving and do not have the restraint of a need to be profitable. The "invisible hand" of Adam Smith applies only to business, not government. Government is certainly more powerful than business and has little need to be accountable for the actions of its politicians and bureaucrats. Government can enforce their rules using handcuffs, batons, guns and threat of jail, business cannot. All politicians need do is convince the majority of voters they will take from the minority who are "wealthy" and give to the majority who are not. Class warfare and hatred for the successful are their most effective election tools, even though they actually do exactly the opposite of what AGW politicians promote. Since 2008 the largest movement ever of money, property, and power from mainly the middle class and poor to the very wealthy has occurred in the US. Of course,

the growing perks for politicians in power are paid by taxpayers and their personal wealth is exempt from criticism. This is especially true for those holding the highest offices or positions. They are the "foxes living in the chicken coop."

In most organizations, the power is expressed from the top, through the leaders, however they rose to power. In families, the structure and the leader are usually decided by birth. In governments the structure and the leader are chosen by many different means. In our democracy we like to think they are chosen by the people in free elections and by the officials we elect. In many countries much the same as in troops of baboons, the leader is usually the biggest, strongest, meanest nasty in the bunch and that ruler rules with the power of death over the subjects. Afghanistan under the Taliban, Germany under Hitler and currently, Iran and other Islamic countries, are classic examples of the latter.

In most of the world, *business* does not have that kind of power. Except for government supported cartels such as OPEC, government operated businesses like the American Postal Service, or even those mega corporations in bed with politicians and buying and controlling their votes. Most businesses have one or a group of owners whose basic purpose is to generate income, a living if you will, for the owner or owners. In precisely the way an individual farmer or worker works to support his family or a performer goes on stage to make a living, the business person seeks to earn income to spend as he sees fit. Each worker, professional, business person, politician, artist, individually or collectively works to enlarge his income or *piece of the pie* and enhance his power (find his place in the pecking order) over other individuals. Millions of years of evolution developed within us this modus operandi along with other strong forces that now threaten our very existence.

A word here about that oft described and accused monster, the military/industrial complex. It's an erratically described, hazy group of assumptive power brokers from the two named groups who supposedly run things from behind the scenes. If there is such a group, the name ignores the most powerful members of the group, politicians. Politicians order the military equipment, decide which organization or company gets the order, and personally gain substantial perks, election financing, and income under, through, and/or around the proverbial table. Why do you suppose there are so many nouveau riche in our Congress? Should you analyze Eisenhower's famous phrase, "Beware the military/industrial complex." you will find that now, the military is controlled and funded by members of Congress, the government. Industry is now mostly controlled by the banks. This means that statement should now read, "Beware the government/banking complex." I couldn't agree more.

Because of the huge number of *businesses* in existence it stands to reason that truly evil or uncaring people are members and even control some of them. But that is also true of the biggest *business* on earth, the US Federal Government. There are many truly evil people in government, both elected and appointed or hired by the bloated government bureaucracy and I'm quite certain the percentage is much larger than in the average business. This is especially true of those who are elected. To call *business* and *the bottom line* the cause of our environmental problems is as ludicrous as to so call *attorneys* or *the Hollywood crowd* or *labor unions* or *government* the cause. The real cause is simply, too many humans.

My bet is that most informed business people are as concerned about the environment as the average informed citizen of most of the world, probably more so, particularly if their principals are long range planners. I see the real reason for the condemnation of *business* in this area as the old *class warfare* theme used by power hungry people to anger the masses against those who have wealth and power, no matter how hard they worked to gain that wealth or how many good jobs they provided in the process. Success often feeds the desire for power and those who have power always accuse others who have power with abusing it. What is the motive of the famous and powerful elite of the academic and entertainment worlds in attacking businesses. These are the same businesses whose profits afforded the means that provided them with fame, wealth, and a platform from which to speak? Now that their wealth and position are secure, they use class warfare to seek admiration of the masses by attacking others with wealth that are not so well known and, in particular, corporations without faces—publicly held companies. It's people with power at war with other people with power and with no real concern for the little guy whatsoever in spite of their many pronouncements of how much they care. If they really cared about the poor, they would donate their own wealth to help them and teach them how to make a living. Instead they exercise their power to donate the wealth of others, not their own. That's political compassion paid for by others.

Back to the epidemic: What is it that drives the engine of destruction that threatens us all? What is the real reason for the shrinking forests, warming atmosphere, dying corals, extinctions of species and other cataclysmic happenings those scientists spoke of in 1992 when they said, "our current practices and activities are altering the world in ways that make life unsustainable?" Why is it these so-called *wise* men couldn't or wouldn't speak the truth about the real cause of the problem? What are the actual *practices* and *activitie*s at the root of the problem? The answer is quite simple. Given the real facts, any twelve-year-old could figure it out. The answer is an epidemic sweeping the earth and destroying in a few years what took millions of years to create. This epidemic is the unconscionable and totally uncontrolled births of many more human babies than can possibly be fed and cared for. Why does no one speak the truth about this?

We hear many expressing the need for the production of more food to feed the coming billions. Where do they propose more wilderness be turned into crop lands? What woodlands do they propose we destroy to feed a billion more wood burning cook stoves? What wild habitat filled with wild creatures do they propose we eliminate to make room for human villages and cities? What idiot penned the following nonsense, "The world population in 1960 was three billion. Today it has doubled to a little more than 6.2 billion. It will double again in less than 50 years." No one dares to say it, but it is physically impossible for our earth to sustain 12 billion humans and at the same time protect and support diverse wildlife. When our ecosystem collapses, none but the tiniest life forms and a very small number of larger ones will remain. Cockroaches, ants, and similar creatures will inherit the earth.

As the epidemic of uncontrolled births rages, most babies are born to poor, ignorant, starving women with little hope for any real relief. As their numbers rise, so, too, do the numbers facing disease, starvation and death as their only future. Why do those proponents of no restrictions on the creation of people, point to the US and secondly the rest of the Developed World as the causes of the suffering? Why do they decry the creature comforts of those of us who have the lowest birthrates? Would they expand the Third World population rates to the entire world thus ensuring an even earlier cataclysm? Apparently they would reduce us all to the common denominator of starvation and disease. The few to initially escape such personal devastation will be the politically powerful leaders and their cadres of lieutenants and enforcers. Unfortunately for them, their reprieve will be short. See *The Lemming Syndrome,* page 76.

It amazes and distresses me that so many of even the most intelligent and educated of us are such unthinking prisoners of the biological instinct to reproduce. Most ten-year olds can understand the math and its horrible results. The quite practical Chinese are the only ones on the globe doing anything about the problem and they are under fire from much of the rest of the world. Talk about ignoring reality. I can understand this from dictators, religious leaders and despots of all kinds who promote population expansion to grow their own following or create soldiers for their armies. They have a purpose and usually do not care who suffers as long as their purpose is fulfilled. They are merely satisfying their own desires without concern about the consequences. To me, they are high among the really evil ones who are destroying our planet. What I can't understand, are the constant and impassioned denials of the reality and imminence of the menace by decent, otherwise intelligent people.

With tremendous publicity on TV and all other media, those who kill or destroy wildlife are condemned as they should be. How about being realistic? For every rhino killed for its horn, for every great cat killed for its skin, for every bear killed for its body parts, for every animal killed for any reason, there are hundreds or thousands denied life by man's continuing expansion into the wild with the resulting habitat destruction. Even the wild habitat of the sea is now being increasingly devastated by over-fishing and bottom-dragging nets. The destruction of mangrove swamps and the loss of other brood areas for sea life merely adds to the problem.

Perhaps the planet can support three billion humans without cataclysmic changes to the biosphere, maybe even twice that many, but certainly no more. The changes mentioned along with other undiscovered menaces, have already signaled we have past the point where unlimited expansion of population can continue safely. There are thousands of indications all over the world that this new cataclysm has begun in addition to those I mentioned. The decline of many amphibian populations all over the world, Saharan sand reaching the Caribbean in increasing amounts, the collapse of the north Atlantic fisheries, the accelerating destruction of rain forests, all are indications of changes wrought by demands of an increasing number of humans on an earth with strictly limited resources.

The extinction of the dodo and the passenger pigeon were but an early indication of what humans are doing on a much larger scale today. Many wild creatures and plants are disappearing almost unnoticed as humanity expands into wild places destroying jungles and forests with their great diversity of life. Wetlands are drained, rivers are diverted, even the floor of the oceans are being changed by bottom dragged nets. Slash and burn, and other destructive agricultural practices destroy wild habitat and the plants and animals that live there. Many of those life forms are unique to the area and will never return. Extinct means gone for ever. It has been said and rightly so, that we are in the midst of the most extensive extinction of life since the great Permian extinction of a quarter million years ago. This is the *Easter Island* effect on a global scale.

We frequently and sometimes enthusiastically address the many symptoms of the problem. Unfortunately it is usually with feeble efforts having little effect in comparison with the real damage being done. Unless we stop this rapidly expanding menace very soon, the entire earth will follow in the way of Easter Island to become a barren relatively lifeless waste with little biodiversity of any life larger than microscopic. Easter Island is a clear example of what happens when humans reproduce without limitation in a limited environment. There they created an ecosystem where all but a few organisms have become suddenly extinct. And this will very likely happen to our earth within the lifetime of a few generations. Historically, these kinds of collapses are sudden and usually unforseen. The signs of the collapse are usually ignored until it is too late. For actual examples, read **Collapse** by Jared Diamond or **Nicholas and Alaexander** by Robert K. Massie.

What must be done to stop this population juggernaut? Zero population growth must be demanded of all people on the earth and enforced. Accomplishing that would at least halt the decimation where it is now. All religions must stop the ridiculous and damning promotion of child production and begin promoting birth control. International law must limit individuals to one child each, or two per couple. There can be no exceptions. That one birthright per person could even be made saleable. It may seem unjust, but a person who didn't want or couldn't have a child, could sell that right on the open market. This could provide a source of income for the very poor while moving many births from the very poor to the educated and successful. Certainly this would eliminate the individual's right to reproduce uncontrolled, but there would still be some choice. We have the technology to implement such a program, but do we have the stomach for it? Considering the alternatives, I see us with no choice at all. If we do not do this soon, the decimation of rain forests, extinction of species and global warming will all become irrelevant. Bare survival will be the only driving force, and soon the planet will be devoid of all large animals except humans and their food animals. I really doubt the human species will last to that point. Some new organism will wipe us out in one single cataclysmic epidemic. New killer organisms are constantly evolving. Life on the microscopic scale evolves and can adapt infinitely quicker than can any large animal.

Until the last several years, I believed the disintegration of civilization caused by overpopulation to be a century or so in the future. The increasing barrage of environmental tragedies; the visible, incremental changes in local environments all over the earth; even the inexorable mathematics, when applied; all provide unmistakable signs of this growing menace. I've shared a few in this talk but there are hundreds more. I now believe this collapse is already beginning and will grow to catastrophic proportions within the next twenty to forty years. During this time the world population will increase by four to six billion while world food production drops drastically. It will be an accelerating collapse with many wars and billions of refugees flooding every productive area of the planet causing a chain reaction of destruction. We in America are not be immune! Look at the invasion of legal and illegal Moslems as well as Hispanics since Kennedy changed the immigration laws. Kennedy promised:

"First, our cities will not be flooded with a million immigrants annually. Under the proposed bill, the present level of immigration remains substantially the same...

"Secondly, the ethnic mix of this country will not be upset... Contrary to the charges in some quarters, [the bill] will not inundate America with immigrants from any one country or area, or the most populated and deprived nations of Africa and Asia...

"In the final analysis, the ethnic pattern of immigration under the proposed measure is not expected to change as sharply as the critics seem to think... The bill will not flood our cities with immigrants. It will not upset the ethnic mix of our society. It will not relax the standards of admission. It will not cause American workers to lose their jobs."

Like most Liberal and all of Kennedy's promises things turned out the opposite. This doesn't even mention the millions of Islamic immigrants, mostly young men and mostly illegal who have entered the US, or the millions of Islamic refugees, also mostly young men who Obama wants to bring into the country. Of course, what would you expect from a Moslem President.

Some time ago I brought the subject of overpopulation to the attention of a popular radio talk show host. I was told it would not be a good subject for discussion. People would not like to hear about it. Apparently people do not want to hear about bad predictions that directly involve them. They want to hear bad news about others and good news about themselves. Bearers of bad news for all are not greeted warmly. Since our news media have become part of the entertainment industry, it has focused on stories involving mostly tragedies, the bizarre and the feel-good aspects of life. It feeds those spectators of tragedy that cause *rubberneck* traffic jams around accidents as people strive for a view of death and destruction from the relative safety of their own vehicle. Shows depicting horrible violence are watched by millions in the remote safety of their own living rooms. Stories about the destruction of the earth by asteroid or comet strikes have been quite common recently. Since it is an extremely remote possibility, we view those stories as entertainment. How would we react should there suddenly exist the certainty of a Delaware sized object striking the earth in ten years? The menace I speak of is real and has similar destructive potential. If we do nothing, it will happen, and more likely sooner than later.

In discussing this with friends and family I find nearly all fall into several groups. To those who believe the apocalypse (the book of Revelations) to be almost upon us and thus are unconcerned about the future, I say, "We have expected the Biblical apocalypse for nearly two thousand years. Suppose it doesn't come for another two thousand, or even a million years?" To those who genuinely believe God will somehow provide a magical solution, I say: "God made us stewards of this world. Are we to abandon that charge and rely on him to provide the solution?" To those who believe really concerned people are working to reduce population and will find a solution soon, or those that believe we can continue to expand our food supply (perhaps by some magic) pushing the real danger of overpopulation centuries into the future, I say, "wake up people! Review your grade school arithmetic. The facts are there. The math is simple and incontrovertible. Population continues to expand exponentially. Food supply has stopped growing. In fact, we may already have past the point of no return. Malthus was actually right only he was a bit off on his timing and too optimistic!"

We've watched the lemmings do it for centuries in limited areas. See *The Lemming Syndrome*, page 76. Humans now have the capability to do the same thing with themselves over the entire planet and we've nowhere to hide! That is the real apocalypse we face and it is now upon us! The growing wave of millions of jobless and hopeless now migrating from poor, starving, and overpopulated areas into mostly Europe and the United States and the growing bloody violence in the Middle East are confirming factors of the growing overpopulation problem. Like lemmings, starving humans are on the move away from overcrowded lands. The results are devastating more and more of our planet.

It has been suggested to me that I have posed many problems but no solutions. I beg to differ! The obvious answer is to control the birthrate and to stop relying on the death rate to control population. The real challenge is how to do it. It is axiomatic that the very first step in solving any problem is to state the problem as accurately as possible with the second step being to suggest a number of solutions. I say that before a problem can be addressed it must be acknowledged that one exists. As long as those in positions of power in the world deny that overpopulation is a real and present danger and the root cause for so many of our ills, very little will be done about it. Where is the cry for population control from the environmental movement? - the Sierra Club? - Greenpeace? - all the others so outspoken about the damage man is doing to the planet? My mission, therefore, is to call attention to the problem, to wave a red flag of imminent danger to all who will listen in hopes it will eventually reach those who can do something to turn this growing tide, even on a tiny scale.

We humans have a penchant for ignoring or denying the existence of many problems we face. Unfortunately, this is one of those inexorable dangers that must be dealt with. It is a real menace and is not going to go away. If by some miracle our planet can sustain five billion more births, what about ten billion, or a hundred billion? As a person who jumps from a plane will eventually reach the surface of the earth, population growth will outstrip livable habitat and food resources! Evidence of civilizations that have destroyed themselves by outstripping those two

necessities abounds. The Aztecs of Central America are but one example. Their civilization grew and flourished with a very high living standard, then nearly disappeared before the Spanish arrived. By then their living standards had been greatly reduced as livable habitat declined and starvation was quite common.

Easter island is denuded of most of the original vegetation, has only chickens and rats remaining of a once diverse fauna, and has less then 5% of the human population it had once. Even the shellfish populations of the coasts were decimated. There are no records of how the population was so vastly reduced. Obviously more people died than were born for a finite period of time. What ensued during that period remains a mystery. One conjectures many horrors.

Sri Lanka will probably be the next island to go the way of Easter. It is currently nearly denuded with less than 5% of its forests remaining. With its agricultural land eroding, its food production declining and its population exploding, Sri Lanka will soon be another microcosm of the rest of the earth. What will probably happen is a major civil war between the two ethnic groups which will accelerate the destruction of even the human habitat.

Madagascar will probably follow and then the continent of Africa where overpopulation has already denuded huge areas in the Sudan and Ethiopia. All over the planet, wilderness is being destroyed and turned into exclusively human use. Wild plants and animals are being destroyed, killed or driven into more remote and inhospitable regions. How long do you think it will be before those game preserves in Africa are overrun by starving humans much more interested in survival than in protecting bio-diversity? Though the timetable is unclear, the next 20 to 40 years I mentioned before are about all we have left if there is not a reversal or at least a substantial reduction in population growth rates.

Whether or not you agree with my timetable, the end result is inevitable. Most certainly we must recognize that there is a problem before any action can be taken to solve it. I hate to end on such a depressing note so let's look for some small glimmer of hope. As I stated before, China has been somewhat successful in stopping population growth, Sadly, they have been condemned by much of the rest of the world for their method. There are a few groups trying to get this message across to the rest of the world with the hope that at least the attitude toward population control can be changed. Religious and cultural groups are an impediment that is extremely resistant to this and they have great power. Somehow, they must be convinced to reverse their attitudes about population controls before the general population will consider any concerted action. In a way I can agree that Armageddon is upon us, but it is not the Armageddon of the Bible. It is of our own making, driven by greed and emotional biases. It is entirely up to us to decide to stop this terrible menace before it destroys us. I only hope this will happen before it is too late! I truly fear we will do nothing effective to prevent the planet from going the way of Easter Island. In the end, all but the tiniest life forms and a few food creatures will survive, the forests will be gone, and man's numbers will plummet. Many SciFi stories have described such an end to humanity. Evidence for such patterns of societal collapse is shown by records of many island ecologies around the world. One accurate assessment and you will realize the earth is just such an island in space.

Will we turn our planet, our island earth into another failed island system, a giant Easter Island, or will we take the steps, realistic steps, that are required to prevent such a catastrophe? I seriously doubt that humans have the wisdom, courage, and rationality to take the steps necessary to prevent such a sorry end.

NOTE: What will most likely happen is an ecological disaster, a devastating and sudden collapse. The entire earth will become an Easter Island, denuded of most life including man. This has already happened numerous times on islands where a single species has consumed its entire food supply and starved to extinction. We are on an island planet. When our prey/food animals are all eaten, Man, a predator, will most likely become his only food source. Who can predict what horrors that will bring? The Earth will very quickly end up populated by rats, ants, and cockroaches as the surviving life forms. Like the Easter Islanders, man collectively is probably too stupid to take the steps necessary to prevent this catastrophe.

❖   ❖   ❖

The following is an excerpt from a report based on the first two references listed at the end. The largest section of this report is highly technical and may not be understood by the majority of laymen who are unfamiliar with the physics or thermodynamics of gasses. Even scientists without an in depth understanding of the complex physics involved may have difficulty understanding some of the realities of the math and physics. This part of the report is not included here as it is beyond the understanding of most people not fluent in the equations of physics. For those interested, the entire report can be read on the Internet at:

http://mysite.du.edu/~etuttle/weather/atmrad.htm

It is unfortunate that politicians and media personalities who have virtually no understanding of the scientific realities are the ones informing the gullible public of the non existent dangers of increased atmospheric carbon dioxide. In fact, the carbon dioxide already added to the atmosphere has had a huge positive effect on agriculture, world wide. Experts in the field who study this claim grain and vegetable production has increased nearly 15% to 40% because of the increased carbon dioxide in the atmosphere. This increase in the food supply has fed upwards of a billion people, a very positive and desirable result.

An Interesting Report:

The Global Warming Debate

There is evidence that the average temperature of the Earth is increasing. *(Depending on the historic period chosen, there is also evidence it is decreasing)* The rate of increase *(or decrease)* is slow, and is certainly not unreasonable. The average temperature has fluctuated rather widely in recent geological history. In fact, it is generally assumed that we are in an interglacial era, and that the temperature changes are less remarkable than if it remained unchanged. The reasons for continental glaciation are still quite unknown, and prediction is not possible. It is somewhat remarkable that permanent ice still persists at polar latitudes and high altitudes, since this does not appear to be typical in geologic history. At present, then, it would be reasonable for the Earth's temperature either to decrease or to increase, since it is at a rather intermediate level, perhaps cooler than normal, so an increase would not be surprising.

The argument current among some scientists, politicians and the general public (not remarkable for geologic knowledge) is that the increase in temperature is caused by carbon dioxide emitted into the atmosphere by human activity, and that restriction of coal burning by electrical utilities, together with some less effective measures, will reduce the carbon dioxide concentration and solve the problem. It is indeed an inconvenient truth that this simple argument is rubbish.

We have noted above that by far the most effective greenhouse gas is water vapor. Some very small increase in its atmospheric concentration, perhaps caused by human activity, would also cause an increased greenhouse effect, and an increase in the Earth's average temperature if the greenhouse effect is indeed responsible for climate. Exactly the same argument can be made for water vapor as for carbon dioxide. *(Except that the effect of water vapor is on the order of several hundreds of times larger than for carbon dioxide.)* For example, burning natural gas produces large quantities of the principal greenhouse gas, while if coal is burned to produce the same amount of heat, only the much less effective carbon dioxide is emitted, turning the usual argument on its head. The atmosphere is no more a closed system for water vapor than it is for carbon dioxide, and what is added may not end up in the atmosphere after all. In fact, agriculture could be responsible for much water vapor, and since agriculture increases at the same rate as population, this would provide an anthropogenic source as well.

Not only is water vapor not mentioned in connection with global warming, neither is the effect of population, except peripherally. If global warming is anthropogenic, then the only means of preventing it would be a significant reduction in human numbers, which seems politically impossible. It is another inconvenient truth that there appears to be no way for human population to be self-limiting until resources are exhausted and starvation does the job. Russia seems to be the only major country expecting a decrease in population (which they are doing all possible to avoid). This is valid even in the carbon dioxide picture. Predictions are now being made for times when the population will certainly exceed the resources, as soon as 2050, when the population will (hypothetically) have doubled once more. How much limitation of carbon dioxide can be realized in this case?

More carbon dioxide and warmer weather are good news for plants (they survive and give us food even with the small amount of carbon dioxide available in the atmosphere). Such conditions are maintained in some actual greenhouses to increase crop yield, but any positive consequences of global warming or increased carbon dioxide are extremely unpopular with the enthusiasts.

None of the proposals for controlling climate can be expected to have any measurable effects whatever, as good as they may be for conservation and efficiency.

End of the quote from the report.

HJ NOTES: The fact is the increase in CO_2 is already responsible for an increase in crop growth of from 15% to 40%. (See pages 35 & 36 for the details) I wonder why the global warmers never mention that? Another reality they will never mention is that there is 10 to 100 times more water vapor in the atmosphere and that its heat effect is from 20 to 200 times more than that of carbon dioxide. In fact, the hourly variation of water vapor in the atmosphere at any given time and place has a larger heat effect than even doubling the amount of carbon dioxide. This system keeps our planet comfortable for life as it has for hundreds of millions of years while the amount of CO_2 in the atmosphere has varied widely.

Fortunately for life on Earth, water vapor has a self regulating phenomena in that when the percentage of water in the atmosphere gets high enough, clouds form. The clouds reflect the suns heat rays so the atmosphere received less heat energy, cools, and the water vapor falls out as rain. This is the natural system that keeps our planet at a liveable temperature. If not for this self regulation, water vapor would cause a runaway heating or "greenhouse" effect and life as we know it would not be able to exist.

The global warmers will say that even the tiny amount of carbon dioxide increase has tipped the

balance creating a warmer earth. Unfortunately for that hypothesis, the self regulating water cycle would completely overwhelm any heating from CO_2 as it does for increased water vapor lowering the air temperature and keeping it in the comfort zone for life. This self regulating balance has kept the earth habitable and conducive to many forms of life for hundreds of millions of years, even with considerably variation in the amount of CO_2 in the atmosphere.

The AGW carbon control proposals do however provide vast amounts of money and power for politicians, governments, and researchers at the government grant money trough. Governments remove billions as taxes from the masses to pour down this black hole of economic nonsense where virtually no one is held accountable for the money. The media, who should be exposing this inconvenient truth are instead complicit in the convenient fraud, convenient for those greedy Washington politicians and their coconspirators world wide that is.

No wonder so many politicians are on the AGW bandwagon. It's one of their gravy trains.

References:

E. W. Hewson and R. W. Longley, *Meteorology Theoretical and Applied* (New York: John Wiley & Sons).

F. A. Berry, Jr., E. Bollay and N. R. Beers, eds., **Handbook of Meteorology** (New York: McGraw-Hill).

Howard Johnson, *Energy, Convenient Solutions* (St Augustine: Senesis Word)

www.ingramcontent.com/pod-product-compliance
Lightning Source LLC
Chambersburg PA
CBHW042049210326
41519CB00052B/181